The HISTORY of the RISE and FALL of the WORLD'S RELIGIONS and their Evolution

YOUNUS SAMADZADA

Copyright © 2021 Younus Samadzada
All rights reserved
First Edition

Fulton Books, Inc.
Meadville, PA

Published by Fulton Books 2021

ISBN 978-1-63710-141-4 (paperback)
ISBN 978-1-63985-127-0 (hardcover)
ISBN 978-1-63710-142-1 (digital)

Printed in the United States of America

I dedicate this book to my two loving daughters,
who have also been my best friends.

CONTENTS

Acknowledgments 7
Introduction: Human Quest for Facts 9
Chapter 1: The Evolution of the World's Religions 15
Chapter 2: The History of Primitive Religions 27
Chapter 3: Religion during the Neolithic Revolution 46
Chapter 4: Hinduism 65
Chapter 5: Zoroastrianism 87
Chapter 6: Judaism 106
Chapter 7: Jainism 132
Chapter 8: Buddhism 145
Chapter 9: Confucianism 165
Chapter 10: Taoism 179
Chapter 11: Shintoism 194
Chapter 12: Christianity 210
Chapter 13: Islam 244
Chapter 14: Sikhism 282
Chapter 15: Baha'i 296
Chapter 16: Newer Religions 308
Chapter 17: Atheism 328
Chapter 18: Ecumenism 338
Conclusion 347
Chart 1 355
Chart 2 357
Schedule 1 359
Bibliography 361
Index 371
Notes 379

ACKNOWLEDGMENTS

I thank my kind friend, Michael Kirby, for his amazing patience and countless critiques on many different subjects and discussions in the past four years while I wrote this book.

I am thankful to my parents, who were open-minded. Although he was a very devoted Muslim, my father never forced any family members to go to the mosque. When I was around thirteen years old, my father passed away.

After his death, I was especially thankful to my beloved brother Ahmed Shah, who was my best friend, brother, father figure, and supporter of my education. This book could not have been written without his kindness and attention to my future.

I would also like to acknowledge my brother-in-law Haleem Kayumi for his encouragement and positive discussions during my search for the truth.

Finally, I am forever grateful to my publisher, Fulton Books, and to my editors, Lee Ann at FirstEditing and everyone at Fulton Books, who tirelessly edit this book.

INTRODUCTION

Human Quest for Facts

I was very young, around age twelve, when I developed a considerable interest in learning religion; not only Islam but also other religions of the world. I had a lot of questions. Who is God? Why does God not show himself and talk to us? What does he look like? What does he sound like? Is God a man or a woman? Why doesn't God have kids and a family?

These interests arose from learning in school. Whenever the teacher's response did not convince me, he was mad and directed me to go and "discuss it with your dad." I talked to family and friends; I was never satisfied. Some of the essential items I learned in school yet never received convincing answers to my questions.

First, in school, our religious teacher taught us to believe in God's foreknowledge and his will. Before humans or world creation, God wrote everything down in the Preserved Tablet (Arabic: Al-Lawh Al-Mahfouz). All that happens from conception to death of humans will come to pass, as written in the Preserved Tablet. In general, if a leaf falls from a tree, God knows it before the fall of the leaf.

As part of the school curriculum, we had to memorize so many verses of the Koran about how nothing in the world happens without God's knowledge. Not only I, but also a few other students questioned the teacher: if all humans' actions are predetermined by God and written in the Preserved Tablet, then humans have no power to change them—and neither can God—because God is never wrong.

Human Quest for Facts

Predestination mentioned in Koran several times that God has decided everything that will happen in Muslim's lives and the world. Free will is not mentioned in Koran and therefore the majority of Muslims use the term *Insha'Allah* (English: if Allah wills).

I was confused how, on the one hand, it says in the Koran don't gamble, don't drink alcohol, and so many other don'ts. If you do them, you will be punished by going to hell. On the other hand, we are predestined based on the Preserved Tablet that we *will* (or won't) do those.

According to the Koran, Muhammad is the last prophet, and therefore, Islam is the final religion on the earth. I was confused and worried about the unbelievers in the East and Southeast Asia. They do not believe in the existence of God. God never told them about the last religion of the world and the last prophet, Muhammad. Are these people going to hell?

Many questions arose. Is the religion of Islam designed by God only for Arabs? Or if God created the entire world, as it is written in the Koran, did he write in the Preserved Tablet *before* the creation of the world that those who are not aware of Islam and God will go to hell?

Second, I learned in school that God is in the sky above us. The sky, or heaven, is divided into seven levels, or seven heavens. God dwells in the last level of the sky, or the seventh heaven, because that level is the best part of heaven. It came to mind that God must be giant, powerful, and he sits at the highest level to see what his believers are doing. I questioned why God does not show himself. Is God afraid of Satan, who is continually changing Muslims' minds to not follow God's directions? Other than Satan, nothing else can hurt him, and he can directly tell the entire world that he is God, and everybody in the world would see and believe in him. The world would be in the proper order.

Third, the school taught us Isra (English: Journey of the Night) or Miraj (English: Muhammad's miraculous ascension from Jerusalem to the seventh heaven). Muhammad was sleeping in the cave of Mount Hira. Gabriel (Arabic: Jibril) came to him and transported him on a weird winged creature called Buraq on the twenty-seventh night of Rajab, the seventh month of the Islamic calendar, in 621 CE to Jerusalem or Al-Aqsa Mosque, the farthest mosque.

Muhammad met all the previous prophets at Al-Aqsa Mosque, gave them prayer, and, from Jerusalem, ascended to the seventh heaven. On the way to heaven, four angels showed Muhammad the four directions. He met God, who told him, "Your people should pray five times a day."

Why didn't God trust Muhammad enough to hand him the entire Koran as a whole?

The most exciting part of this teaching was to pray all night. While praying, the world can turn white like daylight and at that moment, the person who experiences this bright light can get whatever he or she wishes; if they touch anything, it turns to gold.

Let's see and think about Buraq, which transported Muhammad to the seventh heaven to meet God. Buraq is a Persian word meaning black horse. The word "Buraq" being a Persian word yet entered in the Koran is very unsettling—or maybe Muhammad loved Persian horses.

Fourth, the school taught that if Muslims are in a war or going to war with infidels, those who are killed go to paradise and those who kill infidels, the same, they will go to paradise. Their place will be in the highest level of paradise. Why were they encouraging Muslims to kill infidels and give them the best part of paradise as a reward? Paradise consists of several levels—seven or eight levels. The eighth level is called Firdaus, where the most righteous Muslims dwell.

The teacher continued to read several verses of what one can find available in paradise. Every possible food, fruit, flowing streams of drinks, and the most beautiful angels as a mate. He was naming the flowing rivers of pure water, the best and purest milk, the best wine, and purified honey.

Fifth, and the event that caused me to start to research about the existence of God, was the death of my sister and brother-in-law. My sister was a beautiful young girl. At a very young age, she developed juvenile diabetes. Insulin was administered only by a physician in the doctor's office once a day—if insulin was available. Sometimes she could not get insulin for weeks. Before the age of eighteen, she developed internal bleeding, and within a few weeks, she passed away. It was a disaster for the family, and my mother lost all her teeth from grief. Based on Islamic beliefs, she was a perfect person, and God took her back; her place is in heaven.

My question for God was why did you create her in the first place? You took her back and the entire family suffered.

After this event, my second sister married a young man. This brother-in-law was an amiable and kind man; he finished his school successfully with high grades and, in the meantime, memorized the entire Koran and prayed five times a day as a dedicated Muslim. They had two children.

One child was one year old, and the second child was only a few weeks old when my brother-in-law became infected with hepatitis B. Every means of cure was available, yet his condition got worse, and he was under strict quarantine; he could not even see his children before his death.

The question arose, why did God create this man? He was ready to start a life and was taking care of the family, and God wanted this man because he is a good person? Taking this young man severely hurt his children and wife, plus families and friends. Based on the Preserved Tablet, he had to go; nobody can change their *taqdeer*, or predestination.

These two events made me more determined to search for the meaning of religion, God, and his existence. I was reading book after book of all the world's religions and could not find any book to answer my questions. I was at the age of fourteen when I committed to writing a book someday that would describe religion and belief and what constitutes them, from the beginning of human life on earth to the present day.

This book starts from the days of humans searching for food and water as hunter-gatherers for their survival. When a colossal thunderstorm washed out their dwellings and carried their children away, the small tribe called the thunderstorm supernatural, or God, and gave respect to it. When a cobra bit one of them, they were thinking of the cobra as a god. When a strong man of the group was able to hunt an animal, the group called that man a leader. They were happy when an animal was killed for their food and tried to show their gratitude to the newly elected leader. They learned dancing from animals, birds, and trees. Next time when they killed an animal, they expressed their happiness with dance. Soon, dance became a ritual.

In nature, everything is progressing, getting better based on past experiences. As time passed, humanity moved from the hunter-gatherer stage to the next stage, agricultural life. This new stage has a society composed of kings, farmers, workers, fighters, priests, doctors, etc. Also, these societies were better for living, and the stronger societies took advantage and attacked the smaller and weaker societies. This way of life became the norm of human behavior.

This book describes every common religion, with facts from the rise to the fall of that religion. It is suited for inquirers who are very religious, moderately religious, nonreligious agnostic, and students. It can be seen how societies with God and without God moved on toward a better life.

Humans are equipped with many emotions, and I believe the fear emotion is responsible for the creation of religions, beliefs, and virtues.

CHAPTER 1

The Evolution of the World's Religions

Bushmen's trance and healing dance (Copyright Alamy Stock photo).

According to the Oxford English Dictionary, the definition of religion is "the belief in a superhuman controlling power especially a personal God or gods entitled to obedience and worship." This definition will apply to the tenets of a few main religions, like Zoroastrianism, and the three Abrahamic faiths of Judaism, Christianity, and Islam. It does not go along with other religions that do not believe in God or gods, like Jainism, Taoism, Confucianism, or Buddhism. However, this definition applies to the majority of the world's population and is assumed to be the proper definition.

There are many theories as to how religious thought originated, and none were satisfactory, until recently the emergence of evolutionary psychology provided a scientific answer.

After the emergence of evolutionary psychology, a new field came into existence that underlined the genetic influences on thoughts and emotions. The majority of evolutionary psychologists believe that religion is not a direct product of natural selection.[1] Religion is part of culture and cultural evolution. "Culture encompasses religion, food, what we wear, how we wear it, our language, marriage, music, what we believe is right or wrong, how we set at the table, how we greet visitors, how we behave with love ones, and million other things."[2] Culture is the way of life, especially the general customs and beliefs of a particular group of people at a specific time. Cultural evolution is the change of these general customs and beliefs over time.

Both cultural and genetic evolution influence human behavior. In short, culture and genes coevolve. According to English author, ethologist, and evolutionary biologist Richard Dawkins, "Evolutionary psychologists suggest that, just as the eye is an evolved organ for seeing, and the wing an evolved organ for flying, so the brain is a collection of organs (or 'modules') for dealing with a set of specialist data-processing needs."

All living animals have emotions, and these emotions are for their survival. Some of these emotions are fear, anger, sadness, joy, disgust, surprise, trust, adoration, awe, confusion, love, contempt, desire, and distress. Humans have directed some of these emotions, like fear and love, toward religious objects. Evolution enabled humans to experience fear because this emotion has served us well, helping to protect us from danger, injury, and death.

British philosopher, logician, essayist, and social reformer Bertrand Arthur William Russell (1872–1970) said, "The most powerful aspect of Bertrand Russel's critique of religious belief is his claim that religion is based on fear, and that fear breeds cruelty."[3]

How the fear originate religion, when a natural disaster like flood, earthquake, or storm killed their loved ones, immediately that natural disaster gave the expanding mind of the group a Higher Power, and they believed for a while that this natural disaster was a supernatural killer. They started to give respect to that natural disaster and developed some sort of ritual and observance toward that specific natural disaster and asked or prayed, do not kill our loved ones again.[4]

Since the early 1900s, scientists have been studying the nature of fear.[5] In 1980, Caroline and Robert Blanchard, working together at the University of Hawaii, carried out a pioneering study on the natural history of horror. They used a rat in a cage, and the Blanchards found the responses of rats to each type of threat with a clear-cut set of behaviors.

Long before the Blanchard research revealed that the amygdala, an amygdalin group of neurons deep down within the front part of the brain, plays a decisive role in the fear-association response in rats, neuroscientists and brain researchers discovered that the amygdala orchestrates human fear as well. If the amygdala gets injured in humans, the injured person cannot detect danger.

In 2010, Dean Mobbs, a neuroscientist at the Medical Research Council in Cambridge, England, tried to study human fear without sending humans to a tiger-infested area. Dean and his team used a system similar to a Pac-Man video game. They programmed a survival-themed video game that rats play while being in a functional magnetic resonance imaging (fMRI) scanner. Mobbs's study was almost identical to the work of the Blanchards.

Mobbs's research also was in complete agreement with some recent studies of rat neurology. One of the midbrain regions that Mobbs and his associates noticed becoming active in humans when a "predator" was near is an area called the periaqueductal gray region. This region showed higher activity in the people who slam into walls more often, providing further evidence that it plays an essential role in panic.

The amygdala and the periaqueductal gray region are ancient parts of the brain, going back hundreds of millions of years. Our small hominid ancestors probably faced the same kinds of threats as baboons in the present time from leopards, eagles, and other predators. The bottom line is man and beast alike tremble at unexpected things. They fear, but they do not know why they fear. This fear is what created rituals and religion, which is part of cultural evolution.

This chapter will discuss some of the essential elements of cultural evolution and emotions, like dance, trance, music, songs, language, and leadership, and how these emotions contribute to religious evolution.

Dance and Trance

Human beings, or Homo sapiens, are the only species that exists today coming from the species known as hominin. Hominin existed on the earth about six million years ago. In this six million years, many species evolved and became extinct.

The age of Homo sapiens is about 117,000 years to the present time, and this is the first human society called hunter-gatherer. This society collected their food, like wild plants, and pursued large wild animals for their survival. The hunter-gatherer existence occupied about 90 percent of human history.

It is believed that complex spoken language did not evolve until 100,000 years ago and that modern humans are the only ones capable of complex speech. The prior extinct species from Orrorin tugenensis[6] to Homo neanderthalensis did not have the anatomy necessary to produce the full range of sounds that Homo sapiens, or modern humans, make. Communication at the beginning was by weird sounds and signs by the hands and body. Language evolved from these communications.

The historian does not have clear evidence when or how dance started. Infants, from the very beginning, show rhythmic movement with feet and hands, and this indicates that rhythmic movements are in the human gene. Infants move rhythmically in response to music before they can walk, talk, or feed themselves. There is no known culture in human history whose members do not know in some way how to move their bodily selves in rhythmic patterns to the sound of clapping, snapping, stamping, singing, or other musical expressions.

The hunter-gatherers faced natural disasters like floods, earthquakes, and storms, which caused a lot of suffering to them, like killing their loved ones and destroying their food sources. Repeated calamities made them think and search for protection from ancestors, natural objects like the sun, lightning, wind, and other animal species, like snakes. As a result, worship started to come into people's minds, and that was dance and trance to express feelings.

Researchers at the University of California-Santa Cruz were recently surprised to discover that one of their sea lions could dance rhythmically, especially when Earth, Wind & Fire is playing.[7] But the sea lion isn't the only animal that can bust a move. Six entities that can dance, according to scientists, are bees, cockatoos, the peacock spider, dung beetles, and freshwater algae.[8]

Geneticists have also addressed the issue of dance. One such study originated at Hebrew University in Jerusalem, where Professor Richard Ebstein and his team have examined eighty-five dancers.[9]

As with studies in the areas of music, sports, and other talents, the media have been attracted to discussions about the potential role of specific genes.[10]

In the case of the Israeli dancers in an Israeli study, researchers focused on two specific genes that provide code, and they involved transmission of information between nerve cells.

1. First was the serotonin transporter gene, which is one of the brain's neurotransmitters and contributes to spiritual experience among other behavioral traits.
2. The second gene was arginine vasopressin receptor 1a (AVPR1a), which has been mentioned in the media for its reported association with athletes.[11]

In some primitive religions, Australian Aboriginals, African Kung and Yoruba, Andamanese, and North American Indians, dance and trance are their main ritual and worship.[12] In some organized religions, like Hinduism, dance and trance have been part of the indigenous Indian subcontinent to the present time.[13] In the Islamic faith, dance and trance are not allowed, but Sufiism includes it in their rituals.[14]

Given the fact that infants are moving rhythmically in response to music, dance must be an intrinsic property of humans. Evidence from several sources reveals a surprising connection between dance and imitation.[15] As in the classical correspondence problem central to imitation research, dance requires mapping across sensory modalities and the integration of motor outputs with visual and auditory inputs. Recent research in comparative psychology supports this association in that entrainment to a musical beat is almost exclusively observed in animals capable of vocal or motor imitation.[16]

Dance and music are abundant in nature, like the sound and movement of the water and wind. Humans learned and used dance and music as first ritual part of their beliefs.

Archaeologists delivered traces of dance from prehistoric times, such as the thirty-thousand-year-old Bhimbetka rock shelter paintings in India. The Rock Shelters of Bhimbetka are at the base of the Vindhyan Mountains on the southern edge of the central Indian plateau. Within massive sandstone outcrops, above the comparatively dense forest are five arrays of natural rock shelters exhibiting paintings that appear to date from the Mesolithic period right through to the historical period. The cultural traditions of the jewelers of the twenty-one villages adjacent to the site bear a substantial similarity to those represented in the rock paintings.

Music and Song

In nature, there exists natural sounds like wind, waterfalls, stream currents, and birds and other animals that have specific harmonic sounds very similar to songs. Many of these sounds are pleasant and have a calming effect on humans, and humans imitate them.

This imitation can be called music. Music is known to every community in every part of the world.[17]

Music is found in all known cultures, present and past, varying widely between places and times. The entire people of the world, including the most solitary tribal groups, have some form of music; that is why it can be concluded that music is likely to have been present in the ancestral population before the diffusion of humans around the world. As a consequence, music may have been in existence for at least fifty-five thousand years, and the first music probably was invented in Africa and then evolved to become a fundamental part of human life.

A culture's music is influenced by all other aspects of that specific culture, including economic and social organization, climate, and close access to better technology. The ideas and emotions that music expresses, the circumstances in which music is played and listened to, and the behavior toward music composers and players all vary between regions and times. Music, song, dance, and trance were combined in the first human rituals and worship, and were performed in happy times and sad times.[18]

Language

As mentioned earlier, complicated language did not evolve until about one hundred thousand years ago. Modern humans are the ones capable of complicated speech more than their hominin ancestors.

There are several theories about how language evolved. Human communication might have been ignited by involuntary sounds, such as "ouch" or "eek," or by communal activities such as carrying heavy objects or heaving, coordinated by shouts of "yo-he-ho."

Another theory proposes that language evolved from the communication between mother and baby, with the mother repeating the baby's babbling and giving it a meaning. Indeed, in most languages, "mama" or similar "ma" sounds mean mother.

As modern human evolution took a long time, the same is true with language evolution. The very first human communication was most likely gestural. In this period of 100,000 years, many languages evolved and many became extinct. Language becomes extinct through migrations, wars, epidemics, and economic conditions. At present, it is estimated that *more than 7,000 different languages* are spoken around the world.[19] The first written language goes back 5,500 years. Evidence has been found at Tell al-Uhaymir,[20] the site of the ancient Sumerian city of Kish. A plaster cast of the artifact is currently in the collection of the Ashmolean Museum, Oxford, England.

All surviving languages change from generation to generation. Anthropologists talk about the relationship between language and culture. Based on the facts, we can consider language as a part of the culture. Culture is the way of life of a particular group of people, enclosing language, moral and religious beliefs, cuisine, social habits, art, music, dance, and behavior. The community of common language will have a shared history and culture.

Language is part of human intelligence. American psychologist Howard Gardner said that intelligence is multiple and includes, at a minimum, verbal-linguistic, logical-mathematical, visual-spatial, musical-rhythmic, bodily-kinesthetic, interpersonal, intrapersonal, and naturalistic intelligence.[21] Human intelligence varies from person to person.

Language is the best means of communication and will establish a particular group or community. Humans evolved as social animals, which is an integral part of the cooperation of group and community for survival. A group's leader must have excellent language skills, as Howard Gardner defined, to influence others with his speech. During the very beginning of the hunter-gatherer period, a leader had to be brave, strong, younger, masculine, and with the communication skills Howard Gardner defined.

The next section will discuss in detail how evolution equipped humans with the leadership and followership skills.

Leader

In all species, a leader evolves from the leaders-followers relationship to allow and assist the species in sharing information and coordinating group behavior for their survival. Examples include honeybees and chimpanzees. Honeybee leaders show the direction to the food resources by waggle dances, and chimpanzee males initiate hostile group actions against predators and enemies.

The same is true with the leaders-followers relationships in humans. Among a group, one individual who has the psychological, physical, behavioral, and social capital, including communication skills, will emerge as a leader, and the rest will be followers.

The culture and religion of every society changes over time because of past experience and knowledge of the society, and that comes through the influence of individuals and leaders whose intentions can vary from good to evil. As the reader continues further, he/she will determine which leaders or prophets were honest and sincere and which ones were evil, lied, used conspiracy theories, and deceived society for their own advantage.

Justin L. Barrett, who is an American experimental psychologist, for the first time he coined a hypersensitive agency-detecting device or HADD. Neuroscientists and psychologists in the last few years proved that human brains are hardwired to make distinctions in their environment between things that are alive from those that are not alive. To illustrate how HADD works consider the following example. When a hunter-gatherer is sitting and waiting for a large antelope to walk by so that he might kill it for his family's dinner, suddenly he sees the grasses in front of him swishing. At this moment, he takes an action for some unexpected event to save his life. This action he takes is an example of HADD, and he sees the hidden agent in a very simple manner. This perception of a hidden agency in some individual becomes so enormous that other thoughts can't act. This individual is called a conspiracy theorist. There is a little of conspiracy theory in every individual on the earth.[22]

In the recent history of one hundred years, it's worth mentioning a few leaders, like Adolf Hitler, Joseph Stalin, Mao Zedong, Fidel Castro, and cult leader David Koresh. These leaders were composed of religious men and politicians and were able to implant fear in their people's minds. Hitler killed approximately seven million Jews among others. It is estimated that Stalin killed between five to ten millions of his own people although accurate statistics are not available because of the repressive and secretive nature of his rule. Mao Zedong killed more than fifty million. Fidel Castro with the support of the Soviet Union governed with an iron fist from December 2, 1976 to February 24, 2008. The cult leader of the Branch Davidians, David Koresh, influenced people with fake talk and lies. He declared himself an incarnation of Jesus Christ and prophet. He and his followers were killed in the fire inside the Branch Davidian compound by his own actions.

A prime example of an evil leader that we are witnessing in our time is Donald J. Trump, the president of the United States of America from 2017 to 2021. He possesses all the qualities of an evil leader. He is power-hungry; and he is an expert on conspiracy theories, deception, and lies. According to the *Washington Post* and many other news organizations, Trump lied during the four years of his presidency more than thirty thousand times. Before his presidency, to seek political prominence, he was promoting racist conspiracy theories that Barack Obama was not born in the United States of America and other conspiracy theories like the conspiracy theory about 9/11. As soon as Trump became president, he attacked and embarrassing people for fighting for their rights like Colin Kaepernick and the press. Freedom of the press is vital to a democratic society, and he was attacking and humiliating newscasters and referred to the mainstream media as fake news. Trump's unusual strongman leadership style and behavior were brought to the attention of psychologists and experts in the medical community. These experts within the medical community soon labeled him as having antisocial, narcissistic, and paranoid personality disorders.

Two months before the November 3, 2020, presidential election, Trump circulated rumors among his followers that the election would be rigged. He knew that because of his dishonesty, he would lose the election. After he lost, he filed many lawsuits in many state courts to overturn the legitimate election without evidence of fraud. All of these lawsuits were unsuccessful. Also, he ignored the COVID-19 pandemic from its start until the end of his presidency. This pandemic killed more than four hundred thousand Americans and brought the economy to its lowest point since the Great Depression despite Trump having a resurging economy handed to him when he came to power.

After Trump lost the 2020 election, he found himself hopeless. All avenues to overturn the election was closed to him and his close associates in the White House and Congress. He invited his followers, whom he had brainwashed with his lies and deceptions, to the Capitol on January 6, 2021, and injected a conspiracy theory of a stolen election into their minds. On that day, a joint session of Congress presided over by Vice President Mike Pence, as acting president of the senate, was meeting to count the electoral votes and announce the winner as the president of the United States. Trump encouraged and directed Mike Pence, his vice president, to disqualify the electoral college votes even though Mike Pence had no legal authority to do so.

While his followers, encouraged by this cult of personality, gathered in Washington, DC, from different areas of the country, Trump, according to New York Times, told the crowd:

> We're going to walk down to the Capitol, and we're probably not going to be cheering so much for some of them, because you'll never take back our country with weakness. You have to show strength, and you have to be strong.

He further told the crowd:

> When you catch somebody in fraud, you are allowed to go by very different rules. So I hope Mike has the courage to do what he has to do, and I hope he doesn't listen to the RINOs and the stupid people that he's listening to."

Trump made sure his close associates would be protected, he told them." He told them:

> "I also want to thank our thirteen most courageous members of the U.S. Senate, Senator

Ted Cruz, Senator Ron Johnson, Senator Josh Hawley… Senators have stepped up. We want to thank them. I actually think, though, it takes, again, more courage not to step up, and I think a lot of those people are going to find that out. And you better start looking at your leadership, because your leadership has led you down the tubes.

Trump, emboldened by this cult of personality, directed his followers toward the Capitol as they chanted "Hang Mike Pence." His followers carried pipe bombs, plastic zip-tie handcuffs, machine guns, and baseball bats. Trump did not go with them. Instead, he returned to the White House to watch the mob's actions and bloodshed on TV. Trump intended to kill his vice president and most democratic lawmakers to create a Trump dynasty with one party's authoritarian system. Trump's followers entered the Capitol by force and broke windows, doors, and stole property. Some of the stolen property included computer hard drives to sell to the Russians. The mob injured many true heroes of the Capitol police force. One Capitol police officer was even killed during the assault on the Capitol. The Secret Service and Capitol police heroes moved the lawmakers from the House Chamber to a safe location to protect the leaders of our democracy. The lawmakers returned safely to the House Chamber later in the night and completed the counting of the electoral votes. After completing the vote counting, the vice president announced Joe Biden as the winner. Democracy had prevailed.

On the other hand, there were leaders like Mahatma Gandhi who did not seek power and wealth. He sincerely and proudly worked as the leader of the nationalist movement against British rule in India. His philosophy and doctrine of nonviolent protests, civil disobedience, or passive resistance achieved India's social and political progress. He also inspired other leaders like Martin Luther King Jr. and Nelson Mandela. Also, his philosophy helped the world against the oppressive forces of racism and colonialism around the world.

CHAPTER 2

The History of Primitive Religions

Tribe of the prehistoric hunter-gatherers wearing animal skins around a bonfire outside of a cave at night. Portrait of Neanderthal / Homo sapiens family doing pagan religion ritual near fire (Copyright iStock by Getty Images/Gorodenkoff).

No historian has definitively concluded when and where religion started. Humans have been living on this earth for millions of years. The first human was Homo habilis, discovered by paleoanthropologist and archaeologist Louis Seymour Bazett Leakey[23] in 1962 in northern Tanzania. Homo habilis lived on the earth 2.5 million years ago. They were smaller, five feet tall, with long arms. Homo habilis is called a handyman because they used stone tools for chopping, and probably used bone and wood tools too. They were living in small groups, and there is no evidence of any spoken language despite having a reasonably large brain. They must have had some sort of communication to collect food and warn one another of danger and predators, maybe sign language and strange sounds.

Some groups were living along waterways. Perhaps a member of the group was suddenly grabbed by a hippopotamus. From the fear of losing a member of the group, they assumed that the hippopotamus was a superpower, and they paid respect and possibly called the hippopotamus a god.

These gods were in power for thousands of years until the expanding minds of new people learned that wind and hippopotamus were not gods. The newer generation, through experiences and the failure of the old gods, started to search for a new protector or god for their survival.

The notion of gods that arose from fear became part of tribal culture. Later, worshipping other wild animals, stones, and trees by the indigenous tribes of Africa, Australia, Siberia, Southeast Asia, the Pacific Islands, and the Indians of South and North America will be discussed later.

As discussed earlier, fear is an emotion in living animals for survival, and this emotion would become part of tribal culture.

Primitive Religion

As discussed earlier, fear can create the notion of God or a Higher Power in human image for survival. The expanding mind of a human is learning discovery and change. Now we will examine that all the faiths that have evolved through the ages have some ritual related to primitive religions.

Animism: Animism is the religious belief in a supernatural power that organizes and animates the material universe or the attribution of a soul to inanimate objects like rocks, river, plants, and natural phenomena. Animistic belief was researched for the first time by an English anthropologist Sir Edward Burnett Tylor[24] in his work in primitive culture, and the term *animism* is accredited to him.

Animism is part of Shamanism, which is composed of beliefs and practices promoting communication with the supernatural. Animism can be seen in almost all religions to some degree, like Hinduism, Shinto, Buddhism, Paganism, and more.

Animism's doctrine is that every natural thing in the universe has a soul. Animism is a primitive belief, and possibly the origin of it goes back to the beginning of humans' hunter-gatherer phase. As humans moved from primitive cultures toward established cultures, animism accompanied them, and therefore, it can be seen in almost every religion.

Animism is worshipping stone, trees, animals, and ancestors. To this date, some religions still follow this practice.

The Greeks worshipped blocks of stone and scraps of wood that they regarded as the bodies of their deities.[25] Pausanias[26] tells of thirty shapeless stones that were worshipped as gods at Pharae, an ancient town of Messenia, Achaia (a province of the Roman Empire), and tells us that among the Greeks, such gems had in old times received divine honors.

Islam worships the Black Rock in Kaaba, Mecca[27] (see chapter 13).

Australian Aboriginals: The population of Aboriginals split into two groups. The first group are those who inhabited the island before colonization. The second group are the descendants of residents of the Torres Strait Islands in all parts of Australia.

The population was very small and spread thinly over the entire continent. It is conventionally agreed by scholars that Australia is the country where the total Aboriginal population maintained the kind of adaptation hunter-gatherer community until modern times.

The History of Primitive Religions

The ancestors of Aboriginals possibly entered the continent about 45,000 to 50,000 years ago. Some other scientists have claimed that Aboriginals arrived sooner, perhaps as early as 65,000 to 80,000 years ago. By about 35,000 years ago, the entire continent was occupied.[28]

Modern humans arrived in Australia about 80,000 or 50,000 years ago, and they formed the ancestors of contemporary Aboriginal Australians. They included the earliest settlers outside Africa. They had no connection with the outside world, and somehow, they were able to fight off and not allow any other people to enter this continent—until the British settlements began in 1788. In 1788, the populations of Aboriginals and Torres Strait Islanders were more than 770,000; by 1900, they were reduced to 117,000 and, presently, are around 800,000.[29]

The Aborigines had the least knowledge about creation, nature, and religion. No unity of faith, belief, and experience existed among Aboriginal tribes. Their philosophy is known as the dreaming, also called Dreamtime or in their language altjira, altjiranga, or wongar.[30] The Dreamtime is the time long past when the earth was created. Dreamtime is a concept of providing a moral code, living rules, and codes for interacting with the natural environment. They saw their way of life as commanded by the creative acts of the dreaming and was the earlier time of the ancestral beings that emerged from the earth at the time of creation. Their spirituality and laws tangled with the people, creation, and the land, and this formed their sovereignty and culture. Most of them believed that

1. some ghosts or evil spirits are self-created, and these created other ghosts and evil spirits,
2. everything was created by a father who lived among the clouds, and had three sons, and
3. a giant serpent was the cause of everything.

The Aborigines in the south of Australia believed that the sun, moon, and stars were living beings who once inhabited the earth. Some Aboriginal tribes believed that a child's spirit and animal's spirit gain life by entering a female's body. The birth is the outcome of what the creators did, and their power was there at every birth, of children and everything else.

Death and burial rites for the Aborigines were simple, with some variation in each tribe. Their chief burial mode was placing the body of the deceased on a tree and burning it. They would wear white cloth, and widows often shaved their heads.

The deceased's eulogy was in the form of hymns and songs about the dead, depending on the merit of the deceased. The person who had power during life would be much respected, and that person would live forever.

They still do believe in ghosts and evil spirits that come from the dead, and they fear the dead. The demons and evil spirits supposedly move through the air without being seen. Some believe the soul of the deceased ascends to an upper part of the heavens and can come back to visit the survivors. They think of their power to produce rain, wind, storm, thunder, and even have the power to ruin enemies.

The Aborigines mostly believe the spirit and soul of the dead bring discomfort and disease. They try to remove the soul from their homes. The exception is for a person who had power; they want the soul of that person to be in the house, and living the same as if the person was still alive.

The Aboriginals had no temples, priests, and ecclesiastical hierarchy, and practiced together with the entire community. Their rituals were rhythmic physical activity, like singing and dancing, and they danced for many hours until they lost their consciousness and fell into a trance.

They have initiation rites for their survival, like hunting, healing, control of the weather, and, in adolescence, a boy's circumcision with a stone knife.

Aboriginals have several symbols, which include boomerangs and knives for hunting, and a percussion instrument in ceremonies.

African Kung and Yoruba: The Bushmen (Bushman and Kung) are a subgroup within the San people of South Africa who live in the Kalahari Desert. The Yoruba tribe is located in Nigeria and all over West Africa. Both Bushmen and Yoruba have some similarities with Australian Aboriginals. Some of the significant differences between these two tribes will be discussed.

The History of Primitive Religions

Bushmen: The Bushmen did not know that much about God. They believed mainly in the supremacy of one powerful God, and lesser gods or evil power, magic, and homage paid to the spirits of the dead. The Supreme God is who first created himself and then earth, water, plants, animals, and food. He had great power, and protected them from calamity, disease, and bad fortunes. When he was mad, he could send bad fortunes.

The Bushmen were living in an arid, sandy area; water for them was essential to keep them alive. They needed rain, and for this reason, they had a weather doctor. The weather doctor was not sacred to them. If he predicted incorrectly a few times, he was not trusted anymore, and they could put him to death.

They attributed to the lesser god or evil power all evils that happened, like cold, rain, thunder, and lightning. In the case of thunder and lightning, they became violent and shot poisoned arrows and threw rocks toward the thunder.

They held sacred the caddisworm and some species of antelope, like eland. Each one has a purpose. They prayed to the caddisworm for success in hunting, and the eland was a very spiritual animal that became part of their rituals, like a boy's first kill, a girl's puberty, marriage, and the healing dance with trance.

Rituals of marriage, birth, puberty, and death would go as follows.

The marriage ritual is compelling because the woman who marries delivers a future chief or a warrior. Offerings and sacrifices were made to the gods. Not only the bride who marries, but also the groom's family must move to the bride's village. The wedding ceremony would last for several days or weeks.

Birth is another crucial ritual, in particular if male. The male child may grow up to be a chief. Naming a child is a significant ceremony, usually consisting of a given name, followed by the father's name and grandfather's name. Dance, songs, and feasts would follow the naming ceremony.

Puberty is the next ritual for both boys and girls. Boys got circumcisions, and girls got clitoridectomies (a mutilation practice that is losing favor quickly in modern society). Boys had their faces painted before circumcisions, and the ceremony was followed by feasts, dance, and songs.

The death ritual ceremony is the beginning of the next stage of life. The dead person would supposedly go to a place to be with their deceased loved ones. The corpse would be washed, dressed, and placed in the grave with some artifact to help his or her journey. Following the burial would be gift exchanges between family, friends, and neighbors, and animal sacrifices, and feasts.

Bushmen may hold a dance due to a crisis, like easing war between neighboring tribes, or marriage, lack of rain, or hunting. *The Healing Land* is a book by Rupert Isaacson that includes an excellent description of this (p. 123–129). Mr. Isaacson was born in London to a South African mother and a Zimbabwean father. He traveled extensively in Africa, Asia, and North America for the British press.

He notes that "these trance dances and special animals are depicted in the rock art left behind by the Bushmen."

Yoruba: The Yoruba native land is located in West Africa and is one of the three largest ethnic groups of Nigeria; most of them live in the southwestern part of the country. The Yoruba people have a common language and culture and prefer living in cities. They are distinguished from neighboring tribes by language, culture, and religion. They believe in a Supreme Being, in primordial divinities, and spirits that have deified.

The Yoruba people talk about Shango and Oro. The word *Oro* means provocation and fierceness. Oro personifies the executive power or public police. Oro is supposed to haunt the forest and woods around the neighboring towns, and he appears nightly to strike terror and fear. He makes his approach known by a strange roaring noise that is produced by a wood plate twelve inches long and two and a half inches wide and fastened to a long stick. As soon as Oro's voice is heard, women must shut themselves up in their houses and refrain from looking out under the penalty of violent death from Oro.

Shango,[31] also called Chango, is doubtlessly derived from a deified ancestor and a natural force. He was a great warrior and king of the town of Oyo. Shango left Oyo and committed suicide by hanging himself. His followers claimed that he ascended alive into heaven on a chain where he reigns, fights, hunts, and fishes. He is the deity of thunder, lightning, and fire. He favors the rights of hunters, fishermen, and warriors.

Olodumare is the name given to one of the three manifestations of the Highest God in the Yoruba pantheon. Olodumare conveys the responsibility of coordinating the universe and populating the Earth.

Oshun, or Osun, is a deity of the Yoruba. Oshun is associated with water, fertility, purity, love, and sensuality, and she possesses human attributes like spite, jealousy, and vanity. She is one of the favorite deities. She not only gives life, but she also takes it too.

Olorun is the owner of heaven. Olorun brings existence to earth on the order of Olodumare. He is also the conduit between heaven and earth. The pantheon does not get involved in everyday human life. There are no shrines or rituals to him because he is a distant god.

The Yoruba religion seeks to preserve the balance between the humans and the deities and ancestors in heaven.

Andamanese: Andamanese are the dwellers of the Andaman and Nicobar Islands. The Andaman and Nicobar Islands are a group of about three hundred islands in the Bay of Bengal between India to the west, and Myanmar to the north and east. The Andaman Islands are a place of residence to the Andamanese, a group of indigenous peoples whose DNA suggests they are one of the populations derived from Africa via the Arabian Peninsula to Australia. Cultural and isolation studies indicate that the islands may have been inhabited as early as the middle Paleolithic era.

The indigenous Andamanese people appear to have lived on the islands in complete isolation. The Andamanese are called Negritos because they have dark skin and are small, generally less than five feet in height. The Andamanese religion and belief system can be called a form of animism. Worshipping ancestors are the central part of the religious tradition and culture of the Andamanese.

Their social organization consists of small local groups (villages) from forty to fifty people, mostly scattered over the islands, and mainly on the coast, but some of them are in the forest of the interior of the islands. A tribe consisted of several local groups, all speaking one language, each tribe having its own words and its name.

A village would be composed of eight huts, ranged around a central open space, and all of them facing inward toward the center.

This free space is kept bright and clean for dancing, and serves as the village dancing ground.

Each of the single huts is occupied by a family group consisting of a man and his wife with their children and dependents. One shelter is always for a bachelor and a public cooking place. The economic life of the local group is very close to communism—the land is owned in common. The hunting grounds belong to the entire group, and all the members have an equal right to hunt over any part of it.

There exists, nevertheless, absolute private possession of trees. A man of one of the local groupings of the coast may notice a tree in the jungle suitable for a canoe. He will tell others that he has seen such a tree, describing it and its whereabouts. That tree is regarded as his property, even if some years should pass and he has made no use of it, yet another man would not cut it down without first asking the holder to give him the tree.

A hunting party consists of two to five men. Each man conveys his bow and two or three pig arrows. The bow is the only indigenous weapon that they use for fishing and hunting wild pigs—dugong, turtle, and fish are caught with nets and harpoons.

Ceremonial customs are few, and always start with dancing, clapping, and singing. The Andamanese dances are an intensely emotional experience. As the dance continues, songs and clapping with dance intensify, and they keep their strenuous dancing through many hours of the night. The dance also increases the anger against the enemy and builds up the warrior's morale. Men and women always dance separately.

Every boy and girl has to go through the operation of scarification. It begins when the child is very young, and a small portion of the body is completed. This operation is done in intervals until the entire body is wholly scarified. The scarification is done with quartz or glass. The choice of scarification depends on the person who operates. Mostly this operation is performed by a woman. The only reason this custom gets completed is to improve appearance or make the child grow stronger.

A girl reaching her adulthood starts with a ceremony that takes place following her first menstrual discharge. She tells her parents, who cry over her. She must then go and bathe in the sea for an hour alone. After washing, she goes back to her parents' hut or to a special hut that is built for the occasion. She is covered with pandanus leaves and sits with her legs doubled up beneath her, and her arms folded. A piece of bamboo is placed at her back to lean against, as she may not lie down. If she cramps, she may stretch one leg or one arm, but not both arms and both legs at the same time. She may not speak or sleep for twenty-four hours.

This girl sits for three days. Early every morning, she takes a bath in the sea for an hour. After three days, she resumes her life in the camp.

An English social anthropologist, A. R. Radcliff-Brown, believes the first menstrual discharge is supposed to be due to sexual intercourse.[32] The man's breath goes into her nose, and this creates the discharge. It is also believed that if a man was to touch a girl during this period, either during the ceremony or after it, his arms would swell up.

On the occurrence of death, the news quickly spreads through the camp, and all the women collect around the corpse and, sitting down, loudly weep until they are worn out. The women then retire, and the men come and cry over the corpse. All the adult members of the community cover themselves with a wash of common clay smeared evenly over their bodies and limbs. The nearer relatives and intimate friends of the deceased plaster some of the same clay on their heads. They shave the head of the dead, remove the ornaments, and decorate the body. This decoration consists of lines of elegant pattern in white clay alternating with bands of red paint. The higher the estimation of the deceased, the greater is the care lavished upon this last decoration.

Thus, the decorated body is prepared for the funeral. The legs and arms are flexed so that the knees come up under the chin, and the fists rest against the cheeks. A shell is placed in the closed hands for the use of the corpse. A sleeping mat is wrapped around the body, and over this, a few large palm leaves to make a bundle.

Before tying with ropes, relatives of the dead person will take their last farewell by blowing on the face of the corpse. The male relatives and friends proceed to the spot selected for burial, and one man carries the body slung on his back. The women take no part in the actual burial.

A hole is dug three or four feet in depth, the digging done with a digging stick. The body is placed in the hole and the ropes are cut. The body is positioned such that the face is toward the east. Supposedly if the custom is not observed, the sun will not rise, and the world would be left in the dark.

The soil is then replaced, with the men in attendance helping.

Beside the grave, a fire is ignited, and some food contained in a bamboo vessel or a nautilus is left for the corpse. The men then come back to the camp, where the women have been busy loading up everyone's belongings. The camp is deserted; the people move to another camping ground until the period of mourning is over. They may, if they want, return to the deserted campground. No one goes near the grave until the period of mourning is over.

The duration of the period of mourning is not known. In general, it must last long enough for the flesh to decay from the bones. The proceedings at the end of grief are comprised of

- digging up the bones of the dead and
- a dance in which all the mourners participate.

The bones are generally dug up by the man who performed the burial. The men cover themselves with clay and weep over the bones. They wash the bones in the creek or sea and take them to the camp. Women receive them and cry over them in their turn. The skull and jawbone are decorated with red paint and white clay. The head and jawbones of the deceased are preserved for a long time and are worn around the neck of their relatives either in front or behind. These ceremonies transform this dead human into a full spirit.

The power of the spirit may be used to produce both good and evil, according to medicine men, or dreamers. There are three ways in which a man becomes a medicine man.

1. Dying and coming back to life. When a man dies, he becomes a spirit and, therefore, obtains the peculiar powers and quality of a spirit, which he retains if he returns to life.

The History of Primitive Religions

2. If a man stays in the jungle by himself, is confronted by the spirits, and if he shows no fear (if he is afraid, they will kill him), they may keep him with them for two or three days and then let him go. Such a man, on his return, is regarded as a medicine man possessing all the power of a medicine man.
3. A man may become a medicine man by having intercourse with the spirits in his dreams.

The medicine man is believed to be able both to cause and to cure sickness, and to arouse and to dispel a storm. In other words, the medicine man has power for both good and evil, and it can be concluded that the spirits have the same. The medicine man communicates with the spirits in dreams and enlists their aid.

A stranger who dies or is killed is buried without ceremony or cast into the sea. Among the northern tribes, the body of a stranger in the olden days was disposed of by cutting it into pieces and burning it on a fire. The natives believe that by burning all the blood and fat of the dead person goes up to the sky. It will remove all the danger to the living from the dead man.

The burning may indicate that this custom of the burning of slain enemies is possibly the origin of the belief that the Andamanese were cannibals. Marco Polo wrote that the island's inhabitants "are idolaters and are a cruelest and savage race, having heads, eyes, and teeth look like those of the canine species. Their dispositions are brutal, and every person, not being of their nation, whom they can lay their hands upon, they kill and eat."[33]

The Andamanese believe in a religion that can be described as animistic. Among the major beliefs of the Andamanese are those relating to the weather (like wind, rain, rainbows, storms, and thunder) and the seasons. These are under the control of beings named Puluga and Daria. However, all the tribes call on Puluga in every disaster, like in fear of the evil spirits of the jungle, the sea, disease, and ancestors. People try to keep away from any action that might displease Puluga.

Puluga created the sun, moon, earth, all the animals, and she was living on the earth and is now living in the sky. The Andamanese have no ceremonial worship rituals or no idea of hell or heaven.

North American Indian: North American Indians are also called American Indians, aboriginal Americans, Amerindians, Amerind, or First Nation persons. Anthropologists in general are in agreement that the source populations for the migration into the Americas started from a region somewhere east of the Yenisei River around 15,000–7,000 BC. Their DNA suggests that they are of Mongolian, Amur, Japanese, Korean, and Ainu descent.

The North American Indians consist of many tribes, and the resemblance in their religious beliefs is clear. They have faith in altruistic divinities in all nature. They believe in the Great Spirit (God), and did even before the Europeans conquered them and forced them onto reservations by war and broken treaties.

Religious symbolism is significant. Men often revere phallic, aggressive supernatural beings and rain-bringing deities. Religious symbolism is substantial, even in the human interactions of the dance. American Indians believe in an unlimited number of spirits. Communication with the spirits is generally at the hand of medicine men, although anyone is open to "hearing" from the spirit world. The medicine man (or woman) possesses their knowledge from their ancestors, who hand it down from generation to generation.

There are also, to some extent, magicians who profess to possess the power of bringing rain and storms, as well as the gifts of second sight.

The North American Indian tribes never had an actual calendar. They had a single integrated system of denoting days, including a more extended period. A day is a unit; all tribes know it. They did not have names for days. They used a device that was a bundle of sticks of known numbers and daily removed one stick until there were no more stick. To this day, the Hopi Nation in Arizona does not abide by any changing of the clocks forward and backward for daylight savings time, as most of the rest of the world does in spring and fall, and they continue using the bundle of sticks.[34]

Some minor differences in North American tribes are as follows:

The Iroquois Indians: The Iroquois tribal beliefs are slightly different. Generally relocated to upstate New York, they consider one Great Spirit who created the world and adapted all creation to the needs of man. They also believe in an evil spirit, brother of the Great

Spirit. He is eternal, and has some creative power. He created all monsters, poisonous reptiles, and noxious plants. They also recognize inferior beings, both good and evil. They believe the inferior beings are subordinate to the high spirits. To the high spirits, they made offerings. To please (propitiate) the god of the waters, they cast into the streams and lakes tobacco and birds which they had put to death. In honor of the sun and other good spirits, they consume a part of everything they use in the fire.

Most natural objects are in the care of or inhabited by a spirit. Corn, squash, and beans are regarded as a special gift of the high spirits and are each in the care of a separate spirit, having the form of a beautiful female. These three were very fond of one another and loved to dwell together. They are called the three sisters, and to this day many Iroquois and those who know of their practices plant corn, pole beans, and squash together: the beans grow up the corn stalk, and the squash protects the plants from predators.

Iroquois observe six festivals:

1. The Maple Festival, thanking the maples trees for their sweet waters
2. The Planting Festival, invoking the Great Spirit to bless the seed
3. The Strawberry Festival, or first fruits thanksgiving
4. The Green Corn Festival
5. The Harvest Festival
6. The New Year's Festival

When giving thanks to or for various objects of nature, they never burned tobacco. But when invoking or praying to the Great Spirit, they always used the ascending smoke of tobacco, often placed in a peace pipe and shared around a circle of elders.

The Creek Indians: The Creek Indians' religion is polytheistic, with multitudes of good gods or spirits inhabiting some distant region (heaven) where resources are abundant, corn grows all year round, and the springs are never dried up. The evil spirits reside under the earth in a depressing (dismal) swamp full of prickly, scrambling shrubs (briars), and people are half-starved, having no resources in a territory analogous to hell.

Attributed to the evil spirits are famine, drought, floods, and defeat.

They also believe in a penetrating spiritual power that permeates the universe and exists permanently to varying degrees in persons, objects, and places. The creator, called the Master of Breath, heads the pantheon of gods and is followed by the Sun and sacred Fire. Sun is the guardian of the town and the creator's representative.

Other deities include the Moon, Thunder, Corn, and the Four Winds. Creek Indian ceremonies consists of four calendrical ceremonies:

1. Late April or early May is the planting ceremony.
2. Early June is the ceremony and dance for Young Green Corn. It also is the New Year, with the rekindling of the sacred fire and general world renewal.
3. Late August and early October is the harvest ceremony.
4. Late November is the winter ceremony.

The Nootkas Indian or Vancouver Island Indian:[35] The Nootkas, or the Tribes of Vancouver Island, believe in a supernatural deity whose habitation is in the sky. They call him Quabootze, whose nature is little known. When a strong storm begins to rage, the Nootkas climb to the top of their houses, and looking to this great god, they beat drums and chant and call upon his name, asking him to still the violent storm. They fast for the same deity before setting out on a hunt for the same deity. If they succeed, they hold a feast in his honor after their return. This festival is usually held in December, and it was the custom to finish it with human sacrifice. Nowadays, of course, there is no human sacrifice; it's been replaced with a ritual that includes suffering.

In common with other American Indians, the Nootkas have a supernatural teacher and benefactor who came up Nootka Sound long ago in a canoe of copper, with copper paddles. He is said to have told the people that he came from the sky, and that their country would ultimately be destroyed and they would die. But after death, they would rise again and live with him in the sky. In anger, they rose and slew him. But they retain large wooden images representing him. They also believe in many other spirits.

The Dakotas of Minnesota Indian: The Dakotas of Minnesota signify anything they cannot comprehend by the word *wakan*. Whatever is wonderful, mysterious, superhuman, or supernatural is wakan. There is nothing on the earth they do not revere as a god. The only difference they make is that some things are wakan to a higher or lesser degree.

Dakotas believe that all the gods are mortal and propagate their kind. Their Onkteri or god of water resembles the ox. The Onkteri can instantly extend its tail and horns to reach the sky, the seat of their power. The earth is believed to be animated by the spirit of the female Onkteri.

On the other hand, the water and the earth beneath the sea are the abode of the male god. In a religious discussion, they call water grandfather and the earth grandmother.

Among all the Dakota deities, the Onkteri are the most respected. Onkteri sacrifices the feather of female swan, goose, dyed scarlet, white cotton, deerskin, tobacco, dogs, wakan feasts, and dances.

The Wakan men, as medicine men, are believed to have in their bodies spirits, animals, or gods, which give them high power of "suction" and inspiration. They violently suck out diseases from the affected parts of patients. It appears that the medicine men can repel any foe to health until the superior gods order otherwise. The medicine men inflict diseases if they do not get respect, and death is often believed to be the result of the wakan power.

Funeral Customs of American Indians

There are many modes of burial performed by North America Indians. One very common one is placing the dead on scaffolds. The corpses were carefully wrapped in bark and raised on a platform made with transverse pieces of wood lying between the forks of trees.

In some tribes, the body was dressed in the deceased's best attire, painted, oiled, and supplied with bow, shield, pipe, tobacco, and knife for a few days of the journey. A fresh buffalo's skin was tightly wrapped around the body, followed by other robes.

Among the Mandan tribes, when the scaffolds decay, they bury the bones, except the skulls. The heads are bleached and placed in circles of a hundred or more on the prairie, at an equal distance apart, with faces all looking at the center, which they religiously guarded. Every one of the skulls was placed on a bunch of wild sage.

The wife knew the head of her husband and child among the group of skulls. Every single day she'd visit it with a dish of the best-cooked food, which she set before the head at night. She returned for the meal the next morning.

Among the Creek Indians of Georgia, when one member of the family died, the relative buried the corpse about four feet deep in a round hole directly under the cabin or rock on which he died. The corpse was placed in the hole in a sitting posture, with a blanket wrapped around it, and the legs bent under it and tied up together.

If the deceased was a warrior, he would be painted and his pipe, trinkets, and warlike appendages buried with him. The grave was then protected with canes tied up in hoops up to a circle around the top of the hole and then a firm layer of clay sufficient to support the load of a man was added. The relatives howled loudly and mourned for four days.

If the deceased was a man of esteemed character, the family immediately was removed from the house in which he was buried. A new home would be built for the family, with the belief that the place where the bones of their dead were deposited would always be attended by monstrous creatures.

The Comanches packed the deceased upon a horse as soon as he was dead, and took him to the highest hill in the neighborhood, searching for a beautiful place, and then he would be buried privately. The wives of the dead man cut their own arms, legs, and bodies in deep cuts until they became exhausted by the loss of blood. In the old days, the favorite wife was killed, but more recently, only the deceased's horse is killed to carry him to paradise.

The History of Primitive Religions

Dance, Trance, Music, and Song Rituals

The primitive religions live almost the same way today as they did at the past. In every ceremony, ceremonial songs and ritual offerings are followed by group dancing in which visitors and society members participate. For every celebration, there are different dance and songs rituals, like animal dances, the corn dance, stomp dances, the hunting dance, circle dances, men's war dances, buffalo dances, the peace dance, the bull dance, the ghost dance, deer dance, sun dance, and many more. Here we will discuss the sun dance.

Indigenous people believe that unless the sun dance is performed each year, the earth will lose contact with the creative power of the universe, thereby losing its potential to recreate. The sun dance is one of the most famous and spectacular American Indian ritual ceremonies. The sun dance ritual is performed by the tribe's younger men and lasts from four to eight days. Each dancer fasts for a few days before the event to undergo a spiritual purification. They also spend time in a sweat-generating dome.

This dance is a community event, and everybody takes part in its preparation. During preparation, the community elects a sun dance leader to organize the event. The preparation for a larger community takes almost a year. A site is selected, and a large circular area is cleared and cleaned. The tribe's most senior medicine man is responsible for locating a suitable sapling tree. The tree will be cut, trimmed, and brought to the center of the dance circle. A symbolic object of importance to the tribe will be secured between the forks of the tree. The sun dance leader will ritually erect the tree in the middle of the dance area. Once constructed, the tree symbolically connects heaven and earth, where the tribe's guardian spirits reside. From this point, all further prayers and devotions during the event will be directed to this tree.

The next day at sunrise, the sun dance leader, along with the tribe's chiefs and elders, all take their places around the dance site, and the ceremonies start. Throughout the day, drumming, dancing, and traditional songs are sung. During the ceremony, gifts are exchanged, tribal disputes are discussed and solved, and traditional pipes are smoked.

When all the participants have gathered, the drummers start a slow, rhythmic drumbeat, and the sun dance starts. While the participants dance to the drumbeats, they keep their gaze firmly fixed on the sun while reciting prayers and singing praises. The ceremony is grueling, and causes many dancers to collapse and go in trance, which Indians call taking a fall.

When the sun dance is over, the participants' injuries are tended to by a medicine man before they are guided away by their mentors to rest and convalesce.

Sun dance ceremonies typically end with a purification ceremony so tribe members can reenter the world refreshed and regenerated.

Summary

As discussed earlier, the origin of the religion and rituals in primitive society was based on fear, not reason. The emotion of fear is responsible for creating all beliefs and rituals. If a child was playing along a body of water and was grabbed by a hippopotamus or killed by a snake where the child was living, the primitive man found himself helpless, and this caused fear in his mind because of this unexpected event. The hippopotamus or snake became his god.

Even now, some advanced religions worship snakes, hippopotamus, pigs, tiger, cattle, elephants, monkeys, wolves, and dogs.

Primitive man would have been afraid of everything in nature: floods, storms, earthquakes, sun, drought, fire, and much more.

When primitive society hunted and killed an animal, they performed a dance ritual to express their happiness. When somebody died in the community, they had dance rituals to express their sorrow.

CHAPTER 3

Religion during the Neolithic Revolution

Göbekli Tepe is a prehistoric site dating from roughly twelve thousand years ago, in Sanliurfa (Copyright iStock by Getty Images/credit: undefined).

Ancient societies, for about 105,000 years, until the start of the Neolithic revolution or agricultural revolution, collected food for their survival. They were called hunter-gatherers. Hunter-gatherers occupy about 90 percent of human history, and their religions were not part of an institution. They believed in animism, or spirits that controlled their environment and animals around them. They practiced much like the Australian and African Aborigines, Andaman islanders, and North American Indians, which were described in earlier sections.

Hunter-gatherers had a nomadic life searching for food resources. If, for any reason, wild plants and animals were not available, they moved to another location for the search of food resources. Males and females equally looked for food resources, and they were not dependent on one another. Their societies were egalitarian, and the leader was selected based on age in the family or tribe, in which the oldest or the most powerful was the leader. One universal sexual division was prevalent in hunter-gatherer society, that the male is stronger and female weaker. The female did light labor, like gathering, while the more burdensome work, like hunting, was done by males.

At the Neolithic period or agricultural revolution, around 10,000 BC, the population increased, complex languages evolved, and more sophisticated stone technology was created. They learned how to raise cattle, and discovered how to plant. The new learning allowed the communities to settle down with the acceptance of a new way of living in small established collectives, villages, or hamlets in fertile areas with a predictable climate and mostly along the rivers and water.

All the changes with broadening the community, the right of ownership of land, livestock, tools, and on top of all, status, gave them a better life. Simultaneously, the religions and cultures began to change. The powerful grabbed the most significant and fertile land and gradually controlled the villages and hamlets and created a system of government to protect their ownership and power. *Priesthoods were instituted, and hunter-gatherers separated from direct communication with their gods. The priesthood claimed that they were the agents of the gods and had direct contact.*

Religion during the Neolithic Revolution

An excellent example of all these changes is explained in the discovery of Göbekli Tepe, six miles from Urfa, an ancient town in southeastern Turkey. Klaus Schmidt, a German archaeologist, in 1995 discovered this site. He found massive carved stones from a temple, dated to about 12,500 to 10,000 years BC (at the end of the last Ice Age). This temple was designed and positioned by prehistoric people who had not yet developed metal tools. The size of this temple was 49 feet in height and about 980 feet in surface diameter. There are more than 200 pillars in about 20 circles that are currently known through geophysical surveys. Each post has a height of up to 20 feet and weighs up to 10 tons. It is evidence of how the priesthood was influential in everyday life of poor and weak people.

This section will discuss significant religions from the beginning of the Neolithic revolution to the beginning of written language and beyond.

Ancient Egypt

Egyptian civilization started around 5500 BC, surrounded by deserts and dependent on the Nile River, according to some historians based on some written records and artifacts.[36] Hunter-gatherers settled from 5500 to 3100 BC. This period is called the predynastic period, and paved the road for the development of Egyptian civilization in technology, art, and craft, and a very complicated set of religious beliefs and practices.

In 3400 BC, two stable kingdoms were created, one north of the Nile River and another in the south. In 3200 BC, the southern kingdom made the first attempt to conquer the northern territory. A century later, in 3100 BC, the southern kingdom succeeded in unifying the country, and King Menes established the first dynasty.

They built temples to the gods and goddesses and performed rituals to gain their blessings. *Powerful priesthoods were instituted, and those involved in the religious institution controlled the rituals to separate people from direct communication with their gods. The pharaoh was introduced as the gods' representative on the earth.*

People submitted to the rule of the pharaoh. Religion was the strongest and mightiest institution, and as the civilization evolved, the mythology evolved, too, and they had thousands of deities. Tribes had their gods and goddesses, and they worshipped their local deities. As the center of power shifted and another dynasty came into force, the power and prestige of local gods and goddesses grew.

Goddess Maat is the daughter of the Egyptian sun deity Ra and wife of the moon deity Thoth, who represents justice, truth, balance, and morality. Maat covered all aspects of human life and religion, the relationship between Egyptians and their gods, and especially the role of the pharaoh; and also encompassed ethics and politics. The goddess Maat also kept the stars in motion, the seasons changing, and maintained the order of earth and heaven.

The decline and fall of ancient Egypt happened because the central government was corrupt and fighting for power. The Nubians separated and established their kingdom and traded their resources with other countries. The Assyrian ruler Esarhaddon drove the Kushite king Taharqa out of Memphis and destroyed the city. Persian rulers such as Darius invaded and ruled Egypt. Finally, the Roman Empire invaded Egypt and annexed her as one of their provinces.

Creation

Each Egyptian god and goddess had their myths and played their role in maintaining peace and harmony. In general, Egyptians were concerned with divinities of nature, like the sun, the moon, the sky, the earth, the wind, and the stars. They had a lot of stories about deities of nature and their events, often inconsistent.

Some goddesses and gods took part in creation, but some were not connected to the creation legends. Still, they made some contributions, like taking care of people after death, and they became part of this system because of the ethical and political situation. Some gods and goddesses were worshipped more than others.

If the power shifted to a certain locale, such as Hermopolis or Heliopolis, the gods and goddesses of that locale became the divinities of all of Egypt. The eight gods of Hermopolis were called Ogdoad.

Religion during the Neolithic Revolution

These were female and male pairs from primordial chaos and created the world. The nine gods of Heliopolis included the sun god, Ra; his son, Shu; and his daughter-in-law, Tefnut. They were among the creation gods and goddesses, as the practice of creation is repeated every day as the sun rises.

Deities

Ancient Egyptian deities were adopted with specific interests of the theologian for serving the official state religion. There are more than 1,500 of them known by name. The following deities are the most prominent.

Amun: Amun, also called Amen or Amon. He is the great god of Thebes, self-created, and father of the gods. He is shown as a man wearing a pharaoh beard and tall headdress of two vertical ostrich plumes. Amun merged with the sun god, Ra, as Amun-Ra. The temple of Karnak was built in his honor. He is the god of air, and acquired titles of the god of fertility and war. In the Eighteenth Dynasty, Amun was credited with the success of pushing the foreign Hyksos out of Egypt. He also had several merges with other deities: Amun-Ra-Atum, Amun-Ra-Horakhty, Amun-Ra-Montu.

Anubis: Anubis[37] is also known as Anup or Anpu. Anubis is the son of Nephthys and Set, and is the jackal-headed god of Egypt. He guides the dead and is the god of embalming, mummification, and funerals. Anubis presides over mummification with four attendant subordinate divinities. Anubis played a vital role in the afterlife of the Egyptians.

Bes: Bes is also known as Bisu. Bes with his wife, Tauert, were two of the most important domestic gods. They were the protectors of household, mothers, children, and childbirth, and protectors against snakes and terror. Bes's wife is shown as a pregnant hippopotamus and is the goddess of childbirth and fertility.

Hathor: Hathor is the goddess of love, music, dancing, happiness, and fertility. She is the patron god of mining in the Sinai region and abroad. She was worshipped in early dynastic times. Her name means "estate of Horus." She is shown with a horned headdress or as a cow. In her cult center, Dandarah, in Upper Egypt, she was worshipped with Horus. Women aspired to become Hathor in the next world as men aspired to become Osiris.

Horus: Horus was also known as Hor, Har, Her, or Heru. He was the son of Osiris and Isis. He was also husband of the goddess Hathor. He was shown as a falcon whose right eye was the sun, representing power and quintessence. Horus's left eye was the moon, representing healing. Falcon cults have existed from late predynastic times. Horus was the most important of an Egyptian king's title for many dynasties.

Isis: Isis was also known as Aset or Eset.[38] She is shown with a cow's head or with a vulture headdress, and sometimes like a female hippopotamus. She is the protector of a child against snakes and scorpions, and she was goddess of health, fertility, wisdom, marriage, and magic. Many temples were dedicated to Isis.

Osiris: Osiris[39] is also known as Usir. He is the firstborn child of Geb, god of sky, and Nut, goddess of the earth. Osiris is the god of the underworld, fertility, inundation (yearly floods), and the embodiment of the dead and resurrected king. He is shown like a mummified man wearing a white cone-like headdress and curling ostrich feathers with two arms crossed on his breast, holding a crook and flail. Osiris's followers possibly came from Syria and identified their god as Andjety[40] and settled in his city in the Delta. The town was subsequently renamed Busiris.[41] Osiris became a prominent god among the ordinary people of Egypt. During the middle and new kingdom, people saw the possibility that all Egyptians have the chance of an afterlife, not only the pharaoh, which most believed during the Old Kingdom period. Osiris's temple was in Abydos. Osiris's festivals were celebrated annually in various towns throughout Egypt. A central feature of the celebrations during the late period was the construction of the "Osiris garden," a form in the shape of Osiris, filled with soil. The frame was moistened with the water of the Nile and sown with grain. Later, the sprouting grain symbolized the vital strength of Osiris.

Ptah: Ptah,[42] or Phthah, was the creator god of Memphis and maker of things, patron god of craftsmen, especially sculptors and architects. He was shown as a man in mummy form. Ptah was often united with Seker and Osiris to form Ptah-Seker-Osiris. This god combined death, creation, and the afterlife into one god and looked like Osiris.

Memphis was the capital of ancient Egypt after the unification of North and South, and that is why Ptah's cult expanded all over Egypt. Ptah was credited with the establishment and creation of morals, ethics, buildings, town, food, and drink.

Re: Re, or Ra, was the god of the sun, the supreme judge.[43] He is shown as a man with a hawk or falcon head and headdress with a sun disk. He is one of the creator gods. He rose from the primeval ocean of chaos, creating himself, and then created eight more gods and was called the father of the gods. He was also known as the father of humanity and all other creatures. He is linked with other gods like Amen-Ra, Sobek-Ra, Khnum-Ra, and Ra-Horakhty. Ra is one mighty god, starting as early as the Second Dynasty in Egypt, and by the Fifth Dynasty, he was the chief god of Egypt. Ra was believed to travel across the sky during the day. At night, he would die and go through the underworld to be reborn again in the morning. The pharaoh took the title "Son of Ra" to enhance his position among the people.

Set: Set also was called Seth, Sutekh, or Stash. Set is the god of the desert, master of storms, disorder, darkness, warfare, violence, and is a trickster. He is another of the early deities, possibly from Libya. He's shown as a composite figure with a doglike body, slanting eyes, square-tipped ears, and also shown as a man with a doglike head.

Thoth: Thoth is also called Djhuty.[44] He is shown as a man with the head of an ibis or a baboon, animals sacred to him. He seems to have come from the Delta, where he has a lot of connections with all Delta gods. His chief cult center and the temple were in Middle Egypt at Eshmunen of Hermopolis. The original Ogdoad of Hermopolis was called the Souls of Thoth. He was also crucial in the Heliopolitan and Memphite cosmologies, and he is described as the Heart of Ra. Ra said that he created the moon as a reward for Thoth and appointed him as representative in the sky by night.

Thoth is an accountant for the gods and secretary to Ra. Thoth is lord of magic, and taught Isis many spells so that she became a great magician. He is the scribe of the gods, inventor of writing, patron god of writers, the god of wisdom, counting, science, and magic.

The belief in immortality and resurrection was the fabric of the sun worshipper of early periods, as well as the later cult of Osirian. Central to Osirian belief was the hope that when all the body's parts were separated from a person in life, like soul, heart, liver, stomach, lungs, kidneys, intestines, then the person would be restored to life as the god Osiris was after his brother killed, dismembered, and spread him all over the Nile.

Ancient Egypt mythology was not based on rigid codes like modern religions. It was open to change and criticism through cult, usage of worshippers, interpretation, political influence, and time. The search for new symbols was a matter of culture and was continuous. Every new symbol represented a facet of the truth, and the old one was not rejected. As symbols were changed, the interpretation of the myths altered through the centuries.

In ancient Egypt, two prevailing ideas exerted a great deal of influence on the concept of death on other neighboring cultures like Greek, Roman, and Middle Eastern.

1. The first was the Osirian myth of a dying and rising savior god who could confer on devotees the gift of immortality. The afterlife was first sought by the pharaohs, and then by thousands of ordinary people.
2. The second was the idea of afterlife judgment, in which the quality of the dead's life would influence his ultimate fate in the afterlife.

No myth is better than the illustration of this principle than the Osiris myth: Osiris's story completely reversed during history. At the start of Osiris's legend, he ruled Egypt by inheriting the kingship from his ancestors, going back to the creator of the world, Ra. He seemed to be the accessible god, more so than the solar divinities or creator of the world, Ra.

Osiris was a man who suffered on the earth at the hand of his brother and others. He was able to overcome and resurrect the dead through his virtue and his loving wife and son. He introduced the possibility of rebirth for the ancient Egyptians, but also planted life.

Based on Osirian beliefs, Egyptian religion became less dependent on magic and spells against the evil intention of various deities. Under Osirian, an ethical system came in place to lay down a code of conduct for this life, as well as the afterlife. Osirian's influence changed Egyptian religion to an optimistic system that created hope for everybody. This was the reason for the rapid growth of Osiris at the expense of the sun god, Ra.

Many of the solar beliefs were incorporated into the Osiris cult. After the Osiris cult was incorporated into the Heliopolitan system, the sun god became very weak and retained his cult as Amon-Ra, and later was wholly neglected. The Osirian beliefs were not only the Egyptian belief, but also spread outside Egypt in other parts of the world.

Death and Afterlife Rituals

During their entire history, ancient Egyptians spent much of their time thinking of death and inspiring for the afterlife. The existence of so many funerary monuments bears testimony to this obsession. The rites concerning mourning the dead never changed in all of Egypt's history, and are very similar to how people behave toward death today.

In ancient Egyptian mythology, the underworld, or the realm of the dead, was called Duat.[45] Duat was infested with all kinds of predatory animals, evils, lakes of fire, walls of iron, caverns, rivers, and islands. The primary function of Duat is that the deceased's soul goes through to the judgment hall of Osiris.

An elaborate burial ritual is performed for the dead person to prepare them for this transition, and the body of the deceased person is prepared to bear a resemblance to Osiris. (Embalmment will be discussed later.) Priests offered sacrifices and liturgies to Osiris in the presence of the mourners. These sacrifices and rituals were repeated as long as the family was willing to pay for the priests. Also, individual funeral songs were composed.

Most importantly, the relative of the deceased placed a papyrus copy of *The Book of the Dead* or *Ritual of the Dead* with the body. This is a collection of mortuary texts made up of spells or magic formulas to protect and aid the deceased in the passage of several terrifying gates protected by grotesque spirits with human bodies and heads of animals in Duat. The purpose of this dangerous passage is to reach Osiris in the judgment hall.

All around the judgment hall were forty-two judges, human-headed mummies, each with the Feather of Truth on his head. They represent the forty-two provinces of Upper and Lower Egypt. Besides these forty-two representatives, there were the Great Ennead and the Little Ennead. When the deceased passed all these terrifying gates, they would enter the judgment hall, or Hall of the Two Truths or Hall of Double Justice. The trial began with the recitation of the so-called negative confession.

The deceased would respond to each judge and call him by name. No admission of sin was made in the first phase. The second phase of the hearing was presided over by the god of wisdom and reason, Thoth, who was instrumental in persuading the tribunal of the gods Osiris and Horus.

Anubis observed the scale to see if the pans balanced and made sure the heart used no trickery.

At this phase, Anubis or Horus would place in one of the containers an ostrich feather, a symbol for truth. In the other pan, he put the heart, which was considered the seat of intelligence, and indicates the person's actions and conscience. If the scale was equal in weight, it was declared that the innocence of the deceased proved that they should not be thrown to Ammut, a hybrid monster, part-lion, part-crocodile, part-hippopotamus, which was ready to devour the heart of the guilty.

The goddess Maat[46] now dressed the deceased in feathers, like Osiris, as a symbol of justice. Horus brought the dead before Osiris, who announced the verdict.

Religion during the Neolithic Revolution

Embalmment

The embalmment and the preservation of mummies were believed to influence the religious dogma because they seemed to demonstrate that the individual existed after death. The embalmment ritual took about seventy days. The dead person's body was taken away from their home to a particular place, which was called the good house or the place of purification, or the house of gold.

First, the body was cleaned with Nile water. Then an incision was made to take away the liver, stomach, lungs, kidneys, and intestines. These organs were removed and placed in four vessels called canopic jars, filled with palm wine and herbs, and their places in the body were filled with powder of myrrh and other aromatic resins and perfumes.

From the Middle Kingdom onward, the brain was removed through the nostrils, and the cavity was filled with linen or mud. The purpose of supplying the holes was to preserve the integrity of the body intact. It was believed if any part of the body disintegrated, the ka, or personality, would also dissolve.

The heart was called the seat of intelligence and was left in the body in its original place.

The incisions were stitched, and the body placed in potassium nitrate or saltpeter for seventy days, after which it was cleansed, wrapped in cotton bandages, dipped in a gummy substance, and finally coffined and entombed.

Later, a less expensive procedure was used for the general public; oil of cedar was injected into the body, which was then placed in potassium nitrate for seventy days. When the body was pulled out, the fat was removed, along with the fleshy parts of the body. Only skin and bones remained.

A third method, used on the bodies of the indigent, consisted of purging the intestines and covering the body with potassium nitrate for the prescribed period. The body was preserved in precious oils and resins.

Mesopotamia

Mesopotamia is a small region between the Tigris and Euphrates Rivers, which flows to the Persian Gulf. The size of this region is about four hundred miles long and one hundred miles wide. Mesopotamia is a Greek word consisting of two words: *meso*, which means between or in the middle of, and *potami*, which means river. This region of Mesopotamia is called the Cradle of Civilization.

Their influence extended throughout the Middle East, Egypt, and the Mediterranean, and as far as the Indus Valley.

The Neolithic period, or agricultural revolution, took place at about 10,000 BC. Soon after that, by 8000 BC, extensive farming was established all over Mesopotamia, and especially in the south at the inlet to the Persian Gulf, small fishing villages built. The most common species of animals and plants were domesticated. The oldest wheel, dating to around 3500 BC, was discovered in Mesopotamia.

Between 3800 and 3000 BC, Sumerians, from their seat in Uruk,[47] saw enormous growth in population. They developed a system of government like a city-state, along with a trade network, methods of irrigation, construction of residential buildings, and administration buildings with outposts. They changed the mode of operation from the priesthood to that of a kingdom. The civilization in Mesopotamia is responsible for the creation of the four major religions, Judaism, Christianism, Islam, and Zoroastrianism.

Religion in Mesopotamia

The Mesopotamians believed in many gods. The pantheon consisted of more than two thousand deities. There were many spirits controlled by creative gods representing the law and order of the universe. In opposition to the good gods were hostile spirits of darkness and disease. These spirits were controlled by the sorcerer-priests, using exorcism and spells and influencing the gods by their rituals. There were many gods responsible for everything in the world.

The sky and earth were considered creative powers. An (or Anu), the sky god, was the highest god, who played a role in the hymns, mythology, and cults of Mesopotamia. He was the father of all gods and the creator of all good things, especially man. He was also the father of evil spirits and demons, mainly the demoness Lamashtu,[48] who preyed on infants.

In heaven, An assigned functions to other gods; when they were elevated to leadership positions, they received the "An-power." Many temples existed in the name of Anu in different states and cities throughout Mesopotamia, like in Uruk, Der got the title of "city of Anu." Ur-Namma built a magnificent garden and temple for him at Ur, and An had a temple of Babylon in Esagil.

On the earth, An conferred kingship, and his decisions are regarded as final. He represented himself as having a spirit of all moving objects. Anu assumed human form, and their spirit matched men. Anu was symbolized in a headdress with horns, a sign of strength.

The god of deep sea, Ea[49] (whose Sumerian name was Enki), was in the Mesopotamian pantheon of trio gods. He lived in the ocean underneath the earth. Ea was initially the god of the city of Eridu at the mouth of the Euphrates. He came out of the water of the gulf and lived with the men and gave them insight into letters, arts, and science. He taught them to construct houses, temples, create laws, improve their manners, and humanize their lives. He was the god of the wisdom of early Babylonia and was represented as part-man and part-fish.

His association with incantation and magic made him a favorite god for exorcist priests and other deities. With his extreme knowledge of all rituals, he was able to avert and expel evil. His wife was Dav-kina/Damkina, the lady of the earth. His relationship agrees with the Old Chaldaean story of the origin of the world from the deep sea, upon which the earth lay. Through the lady of the earth, the words of Ea were heard by men from the roar of the waves. The offspring of Ea and Dav-kina/Damkina were the gods Marduk/Merodach, Asarluhi, Enbilulu, the goddess Nanše, and the sage Adapa. Marduk was the patron god of the city of Babylon, where his temple, the Etemenanki ziggurat, means the foundation of the heavens on earth located. Ea and Marduk as good gods, and the powers of evils symbolized by a serpent with seven heads and seven tails, there was continual warfare.

Other gods offered Merodach more than fifty names in appreciation. They selected him as their head because he was associated with the god Enlil, who was the head of pantheon and later replaced by Marduk.

Ancient Mesopotamian religion consists of ample myths, epics, hymns, penitential psalms, lamentations, incantations, wisdom literature, and handbooks dealing with rituals and omens. This section will discuss the myth of human creation and flood.

The story of creation comes from the time when humanity was transitioning from hunter-gatherer to the agricultural revolution. Until this time, the gods, rather than man, did the work and bore the loads. The gods complained that the work was too hard and the trouble was excessive. Anunnaki or Anunna (a group of higher deities) made the Igigi or Igigu (a group of gods and goddess) carry the workload sevenfold.

Anu, the father of gods and the king, went to the sky. Enlil, his son and counselor/warrior, took the earth, and the sea was assigned to the farsighted Enki. The gods worked nonstop to dig out the Tigris and Euphrates and cleared the channels. They screeched and blamed one another, and grumbled over the piles of excavated soil. They confronted their lords until complaints reached Anu, and the council of the gods was established.

Then Enki strode in and offered a wise solution to the trouble. Enki addressed the council of the gods, "Why are we blaming them? The labor is burdensome, their work is very hard, the misery is too much, and we kept ignoring complaints. There is an answer to this problem. Call up the goddess, the midwife of the gods, wise Mami. Let her invent a mortal man so that he may carry the yoke, the work of Enlil, the man take the load of the gods!"

The gods' council called up the wise Mami, and they told her to create a mortal that can carry the work of Enlil and let man bear the load of the gods.

Finally, the assembly of gods approved sacrificing one young god and preparing clay. Geshtu-e, a perfect young god who had intelligence, was slaughtered in their assembly, and a ghost came into existence from the god's flesh, and Mami proclaimed it as his living sign. The devil would exist so that none would forget the god. Geshtu-e's blood and flesh were mixed with clay, and the great gods spat spittle upon the clay. Wise Mami, with the help of Enki, created seven males and seven females.

Enlil was pleased with the work of Mami Enki. The newly created human generated rapidly, and Mami gave them new spades, new picks, and led them two by two down to earth to relieve the gods of their labor.

Sixty times 3,600 years had passed, and the land became too wide, the people too many. As the populations grew, the deities sent the Watchers to the Earth. They took for themselves beautiful wives of the daughters of men and begat children of excellent reputation. The holy daughters and sons became the leader of the land. The land became as noisy as a bellowing bull. God grew restless as the noise of humankind became too much, and Enlil burst in, "I am losing sleep!" Enlil organized his assembly of gods and expressed his frustration with mortals for lack of rest to the great gods and ordered them to sicken humankind with headache.

In those days, in the city of Shuruppak, there was a pious man named Atrahasis (Sumerian name Ziusudra). He was the son of the deity, leader (king) of the people. Atrahasis's ear was open to his beloved god, Enki. He was speaking to his god, and his god was talking to him. Wise Atrahasis told Enki, "People are suffering from illnesses. How long the gods make us suffer? You created us, and you should put stop on this unholy sickness." Enki listened to his request then made his voice heard, speaking to his servant Atrahasis. "Call your elders and tell them to uprise against your gods and let them make loud noises in the land and do not revere them for what they are doing to humanity. Go search out the door of the god of the dead, Namtar. Bring him a baked loaf of bread! He may be shamed and wipe away his punishing hand."

Wise Atrahasis listened to his lord Enki and called on elders of the city to gather to his door. He told them what Enki told him. The elders listened to Atrahasis. The elders built a temple to honor the god of the dead in the city. They made loud noise in the land, they did not revere the god and goddess, and they brought a baked loaf of bread to the god of death's presence—the god of death was shamed by the elder's offering and wiped away his punishing hand.

This time the land became very noisy, and Enlil was furious once again; he called on his assembly and talked to the gods his sons that the diseases did not diminish them. They were more than ever before. "I have become restless at their noise. I cannot sleep from their daily activity and noise." He directed the gods to cut off the food on humanity, and the god of rain and thunder (Ishkur)[50] made rain scarce and more calamities to the people.

Atrahasis listened to Enlil's order. They cut rain, food, and all supplies to the people and did all that was ordered. That continued for years. The land did not produce vegetation, the city storehouses depleted, the people's looks were changed by starvation, parents served their children for food, and people stayed alive by taking the life of others. Atrahasis again approached his lord Enki. He spoke to him, "Oh my lord, how long must the people suffer at the hands of the gods? Why the god of rain and flood ceased his rain and fertile flood? People are eating each other, and nobody is safe." Enki listened to his server Atrahasis and directed him to do the same as he did with the god of the dead, Namtar. Wise Atrahasis listened to his lord Enki and called on elders of the city to gather to his door. He told them what Enki told him. The elders listened to Atrahasis. The elders built a temple to honor the god of rain and thunder (Ishkur) in the city. They made offerings to him, and the presence of the people shamed the god of rain and thunder. He withdrew his hand from the land. The drought left the earth. There was plenty of food, and the area became very noisy again. The gods went back to their usual offerings.

Now Enlil was furious with the Igigi (the group of gods and goddesses). Enlil again called on his assembly and addressed the gods, his sons. He made clear that Enki was to release the product for the people; he has corrupted humankind and has instead bestowed upon humanity. Enki fetched for Enlil in the presence of the gods and goddesses. Enlil and Enki were furious with each other. Finally, the warrior Enlil addressed Enki, "You give the people forbidden knowledge, you were given them wisdom, and you betrayed the gods and goddesses by showing them to shame the gods. Your creations have despoiled the earth. You created man by slaughtering a god with his intelligence, and you imposed your loads upon man. Now it is in your power to create a flood to inundate the earth to wipe out life as your punishment." Enki responded and spoke to his brother gods, "Why should I use my power against my people? Could I create a flood to kill my people? That is Enlil's kind of work! Let his envoys march ahead. Let god of thunder Ninurta march and make the weirs overflow and inundate the earth." The assembly of gods and goddesses listened to his speech, but they did not listen to his plea. The gods forced him to swear the oath, and Enlil performed a lousy deed to humankind.

Wise Atrahasis looked for his lord Enki day and night, but he could not find him. Enki came to Atrahasis's dream, and instructed to go to the temple, put his ear to the wall, and listened to his god. Atrahasis went to the temple and placed his ear to the reed wall. Enki spoke to his servant Atrahasis through the temple wall and gave the message: "Make sure you listen to me regularly. Dismantle the house of God, build a boat with its material, make the upper deck and lower deck, roof it like Abzu so that the sun cannot see inside it! Leave all your belongings and put aboard the seed of all living things, a wealth of birds, and a hamper of fish. The bitumen and tackle must be robust to give strength to the boat. The sand needed for flood should be seven nights' worth."

Wise Atrahasis gathered up the elders at his door. He told them that Enki and Enlil have become angry with each other. "I can no longer stay in Shuruppak; I cannot remain in Enlil's land. I must go to Abzu and stay with my beloved god. I must construct a boat to take me there. That is what he told me." The elders heard Atrahasis's speech, and they called the carpenters and all types of workers to his aid. Everything needed for building the boat was brought to him, and the boat was built according to the plan.

All manners of life were loaded aboard the boat. Atrahasis selected the best of all species like cattle, wild beasts, wild animals, and all kinds of birds and placed them on the boat. He put all his family and friends aboard the boat. He invited all his people for a feast. They were celebrating, eating, and drinking, but Atrahasis was going in and out and placing the decks of the boat. Bitumen was brought to him, and he sealed the door with it.

Ishkur, the god of rain and thunder, roared from the black clouds. The winds were raging. Everything light turned into darkness. Anzu, the god of storm and wind, was ripping up at the sky with his talons. The bolt of Abzu broke open, and the flood came out and went against the living things like an army. The flood roared like a tiger; the winds howled. There was no sun, and the darkness was absolute. No man could see anything. The earth was inundated with the power and noise of the flood. Even the gods were afraid of the torrent. They withdraw to heaven, where they cowered.

In heaven, An erupted in a furious rage. He called the gods his son before him. As for Nintu (Mami), the Great Mistress, her lips became encrusted with rime. The gods were afraid of the deluge. They cowered like dogs crouched by a wall. The goddess watched and grieved. Midwife of the gods wise Mami said, "Let daylight return. Let daylight return and shine forth! Enlil was strong enough to give an immoral order. He ought to have canceled that order! What was An's intention as a decision maker?" She was moaning and crying with grief.

For seven days and seven nights, the flash flood, the storm, and the flood came on. Were the bodies clogging the rivers like reeds? When the seven days passed, the flow ceased, silence fell upon the earth. The sea became calm, and the floodplain became flat like a roof. The sun came out and began to shine. When Atrahasis heard the silence, he opened a window in the boat and saw that the sun was shining. He released a dove. The dove came back, for no perching place was visible to it. He took a raven and released it from the boat. The raven never returned. Slowly the water receded, and the boat of Atrahasis came to rest atop on Mount Nimush. Atrahasis put down the door of his boat and let out all his kith and kin and all the animals and his people within the boat. Atrahasis made a sacrifice of thanks to Enki, his beloved god. And there he made an offering for the gods, provided food for the gods. The gods gathered like flies over the offerings. The gods have eaten food like animals. Mami got up and blamed them all. "An and Enlil did not deliberate, but rather sent the flood to eliminate people. You agreed to the destruction. Now the bright faces of our children are dark forever!" Enki looked upon the bodies of his children, which were scattered all over the land. Enki was crying in sorrow for what Enlil had forced him to do.

It was then that Enlil and An arrived. Enlil saw the boat and smelled the sacrifices; he was furious. He called the divine assembly to order. "We, the Anunna, all of us, agreed together on oath. No form of life should have escaped! How did any man survive the catastrophe?" An made his voice heard and spoke to the warrior Enlil, "Who but Enki would do this? He made sure that the reed wall disclosed the order." Enki, the wise, spoke, "I did it, in defiance of you! I made sure life was preserved on the earth!"

Enki's speech humbled the gods. All gods wept and were filled with regret. Mami, the midwife of the gods, cried, "Why have I spoken such evil in the gods' assembly? I gave birth to them. They are my people." Enki and Enlil came to a compromise.

Enki called upon the womb goddess Mami. "You are the womb goddess who decrees destinies. You have created the destinies of the people. You have created the destinies of the gods, and whatever you say shall be made so. I propose a covenant, a bond of heaven and earth, and make Enlil swear to it just as he made me swear to his oath."

The womb goddess Mami made a pattern of flour upon the ground and heaped up a mound and placed it on the center. She commanded Enlil to stand upon this holy mound. Enlil stood upon the mound and pledged to the covenant known as the bond of heaven and earth. Never again harm the people. Never again allow the Anunnaki (the group of deities) to marry with the children of humankind. And so the years passed, and humanity flourished, and the gods were made happy by the people of the land.

CHAPTER 4

Hinduism

The frescoes are erotic inside the temples of the Western group, including Visvanatha-Khajuraho, Madhya Pradesh, India, UNESCO heritage (Copyright iStock by Getty Images/credit: Konstantin Litvinov and KatikaM).

Hinduism is one of the dominant world religions. It originated on the Indian subcontinent and is comprised of several systems of belief, ritual, and philosophy. The term *Hinduism* is relatively new. The first time the word *Hinduism* was used by British writers was in the nineteenth century. Hinduism became known as distinctive to the subcontinent of India, with the publishing of books such as *Hinduism* by Sir Monier Monier-Williams, the outstanding Oxford scholar and author of a dominant Sanskrit dictionary.

The term *Hindu* is derived from Sindhu. Sindhu is a Sanskrit word (an ancient Indic language of India originating from Indo-Aryan language) and is the name of a river in the northwest of India. The inhabitants of the region used the name of this river for the land and its people.

Hinduism is the world's fourth largest religion based on the number of adherents, with close to 1.1 billion based on 2015 statistics from the Pew Research Center. Hinduism is the main religion in India, Nepal, and Bali.

Hinduism, unlike other religions, has no single founder, no single scripture, and no agreed system of teachings. Throughout its long history, there have been many essential figures writing numerous holy books and teaching different philosophies. Therefore, writers often refer to Hinduism as a way of life, or "a whole family of religious traditions deeply rooted in the Indian subcontinent"[51] rather than a single religion. Some scholars believe Hinduism is a polytheistic religion because the Hindus worship many gods and goddesses. Hinduism can be thought of as both polytheistic and monotheistic. Just recently it can be seen as henotheistic (kathenotheist)—those who worship one God while not denying the validity of other gods.

History of Hinduism and Date of Founding

Hinduism may have had roots in the civilization that developed long before writing was discovered. The Bhimbetka rock shelters, from a Mesolithic site dating from 30,000 BC, near present-day Bhopal in the Vindhya Mountain in the province of Madhya, are part of an archaeological site that spans from the prehistoric period to the historic period. These remarkable paintings represent several animals that have been identified as leopards, tigers, antelope, fish, rhinoceroses, frogs, lizards, birds, and unicorn—some of the animal paintings found in the Indus Valley civilization.

The Indus Valley civilization, or Harappa civilization (named after Harappa and Mohenjo-Daro, chief cities along the Indus Valley), in the westernmost part of what is now Pakistan dates to as early as 7000 BC. This civilization reached its height from about 2300 to 2000 BC and extended to more than a thousand sites, at which point it encompassed over 750,000 square miles.

In contrast, as many as 40,000 people once lived and traded with Mesopotamia, Central Asia, perhaps even Ancient Egypt. This culture shows evidence of what may have been a cult of a goddess and a bull—the female figurines commonly used for worship. The bull figurines were used on many steatite seals. These small figurines were found in all parts of India and may have come from pre-Vedic civilizations.

There is no straightforward indication of how the Indus Valley civilization around 1800 BC came to an end, possibly due to drought, earthquake, a changed course of the Indus River, massive deforestation, disease, invasion, or flood. Whatever the cause, a great civilization was lost, and possibly someday, archeologists will uncover all the facts.

The Indus cities may have been destroyed by the flood that inspired the myth of the great flood described in the Shatapatha Brahmana.[52] It has a close similarity to the Mesopotamian story of creation and flood.

The story of Manu (Hinduism calls the first man Manu, the legendary author of the Manu Smriti or Laws of Manu) combines the story of the Hebrew Bible figures of Noah, who preserved life from extinction in a significant inundation, and the first man, Adam. The Shatapatha Brahmana describes how Manu was warned by a fish that a flood would destroy all of humanity. He built a boat, as the fish advised him. When the wave came, he tied the boat to the fish's horn and safely steered to a place on a mountaintop. When the flood receded, Manu was the only human survivor. He made a sacrifice to the gods, pouring offerings of sour milk and butter into the water. The following year, a woman was born from the water. She called herself the daughter of Manu. These two later became the ancestors of a new human race to replenish the earth.

So, in conclusion, the fading and disappearance of Hindus Valley civilization, the Vedic period, came to a close around the same time.

The Vedic period was between 2000 and 1500 BC. The culture of the Indus Valley civilization faded out, and the religion of Vedas faded in. Both cultures may have been a prequel to Hinduism. Although written language existed in India long ago, none of the Vedas information is found in writing.

The Rig-Veda hymns were preserved orally. The hymns composed in the northwestern region of the Indian subcontinent were in an ancient form of Sanskrit. The oldest collection is called Rig-Veda ("Knowledge of the Hymns of Praise"). The Rig-Veda consists of 1,028 hymns, often called mantras ("incantations"), distributed throughout ten books, of which the last and the first are the most up-to-date.

The Vedic texts are a collection of four Vedas: Rig-Veda ("Knowledge of the Hymns of Praise") for recitation, Yajur-Veda ("Knowledge of the Sacrifice") for liturgy, Sama-Veda ("Knowledge of the Melodies and Songs") for chanting, and Atharva-Veda ("Knowledge of the Magic Formulas"), named after a group of priests.

A Vedic text states the code for the recital. "A pupil should not recite the Veda after he has eaten meat, seen blood, seen a dead body, had sexual intercourse, or engaged in writing." It was also forbidden for certain other groups, like infidels and unbelievers, to learn Vedas, as well as pariahs (a member of the low caste) and women.

The origin of the Vedas can be traced from 1500 to 1000 BC, when the Aryans (Indo-European), whose name came from Sanskrit for "noble," invaded India. Both the Sanskrit language and the Vedic religion foundational to Hinduism are attributable to the Aryans.

During the next a few centuries (800–600 BC), each Veda was supplemented by a body of prose writing of later date called Brahmanas. These additions explain the ceremonial applications of the texts, the origin, and the importance of the sacrificial rites of the composed Vedas. An addition of appendices (600 BC) called the Aranyakas and the Upanishads (700–500 BC) explains the more difficult rituals and speculates on humanity's relation to the nature of the universe.

Between 600 and 200 BC, the Vedic religion gradually evolved into Hinduism. The Rig-Veda became the most sacred literature of Hinduism. The Hinduism tradition divided the sacred text into two categories: Shruti ("What Is Heard"), divine revelation conveyed directly from God; and Smriti ("What Is Remembered"), sacred scriptures authored by enlightened sages. Shruti, except the Upanishad and a few hymns of Rig-Veda, is very little known to the modern Hindu. The Smriti texts remain incredibly influential to modern Hindus.

Between 500 BC and 500 CE is the so-called Epic, Puranic, and Classical Age. Hinduism was in a transition period from the religion of Veda to a form based on the worship of new deities, new philosophy, new theology, and new worship systems like the composition of original texts. The Dharma Sutras and Shastras, the idea of dharma (law, duty, truth), which is central to Hinduism, were expressed in a group of texts. The Dharma Sutras recognize the three sources of dharma: revelation (that is the Veda), tradition (Smrti or Smriti), and good custom. The two epics became known as the Mahabharata and the Ramayana, and subsequently, the Puranas, containing many of the stories that are still popular today. The famous Bhagavad Gita is part of the Mahabharata.

The Vedic fire sacrifice was minimized during this period with the development of devotional worship (puja) to images of deities in temples. The rise of the Gupta Empire (320–500 CE) saw the growth of the great traditions of Vaishnavism (which focused on Vishnu), Shaivism (which concentrates on Shiva), and Shaktism (which focus on Devi).

During this period, we can recognize numerous elements in present-day Hinduism, such as bhakti (devotion) and temple worship. This period saw the growth of poetic literature. These texts were first composed in Sanskrit.

Between 500 and 1500 CE is called the medieval period. At the very start of this period, there was a rise of devotion (bhakti) to the major deities, especially Vishnu, Shiva, and Devi. A beautiful granite temple covered with sculptures was built in 1000 CE and dedicated to Sri Brihadisvara (Shiva) in Thanjavur, state of Tamil Nadu. Soon after the collapse of the Gupta Empire, regional kingdoms developed and patronized different religions. Among them is the Cholas Dynasty in the south, which supported Shaivism.

In this period, many magnificent regional temples were built, such as the Shiva temple in Chidambaram, Thanjavur in Tamil Nadu, and Jagannatha in Puri, Orissa. All of these temples had a primary deity installed there, and they became centers of political and religious power.

Poet-Saints and Gurus

The time was ripe during this period for not only religious literature to evolve in Sanskrit, but also in a lot of regional languages, like Tamil. Some poet-saints recorded their religious sentiments. The most prominent are the twelve Vaishnava Alvars (sixth to ninth centuries), including one eminent female poet-saint named Andal and the sixty-three Shaiva Nayanars (eighth to tenth centuries).

Later, teachers and critical thinkers (gurus) consolidated these teachings. They formulated new theologies, perpetuated by their disciples. For example, Shankara (788–820 CE) traveled widely, defeated scholars of the unorthodox movements of Jainism and Buddhism, and established prominent positions of learning all over India. He reestablished the dominance of the Vedic canon, propagated Advaita (monism), and laid the basis for further progress of the tradition known as the Vedanta.

The neighbors from northwest of the Sindhu River had a significant influence on the history of Hinduism, from the agricultural revolution up until 1857 CE.

Between 1500 and 1757 CE is called the premodern period. At this period, the Hindu tradition was in the process of development, mostly in the south of India. At the same time, Islam was rising in the north of India as a religious and political force. The Arab invaders reached the Indian shores around the eighth century, and their army conquered the northwest provinces.

Muslim political power started with the Ghaznavid and Ghorid Empires around 977 CE and continued with other Persian empires until the rise of the Mogul Empire (1526 CE). Akbar the Great (1556–1605 CE) controlled the entire Indian subcontinent, and he was a liberal emperor and let Hindus carry on freely. However, his great-grandson, Aurangzeb (1658–1707 CE), destroyed many temples and restricted Hindu practice. He was imposing Islam practice on them.

This period witnessed further developments in devotional religion, or bhakti. The Sant tradition in the north, mainly in Maharashtra and the Panjab, conveyed devotion in poetry to both a god with qualities (saguna) and to a god without conditions (nirguna), such as parental love of his followers. The Sant tradition merges elements of bhakti, meditation or yoga, and Islamic mysticism. Even at present, the poetry of Princess Mirabai and other saints, such as Tukaram, Surdas, and Dadu, are well-liked.

British Period

Between 1757 and 1947 CE is called the British period. In early 1600 CE, the East India Company was dealing with the Mogul Empire for trading. In early 1700 CE, the Mogul Empire collapsed due to invaders, like Afghans. In 1757, after the Battle of Plassey, which Robert Clive won, the British reigned in India.

At the beginning of the occupation, the British did not interfere with the religion and culture of the Indian people. They allowed the Hindus to practice their religion without constraint. However, later, conditions changed, and the Christian missionaries arrived, preaching Christianity. The first scholars stepped in; they were sympathetic, but often motivated by a desire to Westernize the local population. Chairs of Indology were established in Oxford and other universities in Europe.

Hindu Reformers

In the nineteenth century, some Hindu reformers changed the structure of Hinduism, such as adjusting Hinduism to the modern period, triggering changes in Indian society. From 1772 to 1833, Raja Ram Mohan Roy presented Hinduism as a rational, ethical religion and founded an organization named the Brahmo Samaj to promote these ideas.

Another reformer, Dayananda Saraswati (1824–1883), was the founder of Aurangzeb, a Hindu reform movement of the Vedic dharma. He advocated a return to the Vedic religion, which emphasized an eternal, omnipotent, and impersonal God. He was the first to call for Swaraj, "India for Indians." He denounced the idolatry and ritualistic worship prevalent in Hinduism.

Both reformers desired to rid Hinduism of what they regarded as delusion. These groups were instrumental in scattering the seeds of Indian nationalism and Hindu missionary movements that later journeyed to the West.

Two more essential figures were Swami Vivekananda (1863–1902) and Paramahamsa Ramakrishna (1836–1886), who declared the unity of all religions. Saraswati's disciple Vivekananda developed his ideas and linked them to a political vision of a united India. Swami Vivekananda represented Hinduism at an international religion congress, held in 1893 in Chicago, USA. Vivekananda demonstrated India as a tolerant society that allows different sects to live together under one roof of Hinduism and as a society. He also accepted the people of other religions. He began his speech by referring to other delegates as "brothers and sisters," and so proving his point that all human race was one big family.

In 1920, Mohandas Karamchand Gandhi (1869–1948) returned from South Africa after fighting a battle against apartheid. Vivekananda's ideas were further developed by Gandhi, who launched the noncooperation movement. He was a key contributor to establishing an independent India. Later, he was called Mahatma (great soul). Gandhi, who was a holy man and a politician, is probably the best-known Indian of the twentieth century. He helped work out independence, but was disappointed by the partition of his country. He was assassinated in 1948.

Gandhi drew much of his strength and judgment from the Hindu teachings, such as the concept of ahimsa (nonviolence), and propounded a nationalism that was broad-minded and generous.

Hindutva

From 1500 BC, India was invaded by Aryans and Persians. Especially since the rise of Islam, the Persian Muslims, and later Christians, called the people of India who lived there infidels. During the opposition to colonial rule, the term *Hindu* became charged with cultural and political meaning.

Hindu nationalists have used the geographical association of the word to equate India with Hinduism. By doing this, they exclude Muslims and Christians from their right to thrive in India. A central idea was Hindutva (Hindu-ness), coined by V. D. Savarkar to refer to a sociopolitical force that could unite Hindus against "threatening others."[53]

This ideal has been embraced and developed by cultural organizations such as the RSS (Rashtriya Svayam-Sevak Sangh) and VHP (Vishva Hindu Parishad) and received political expression in the BJP (Bharatiya Janata Party). These sectarian ideas continued after independence.

When India partitioned in 1947, the bloodshed that resulted reinforced nationalistic tendencies and, specifically, notions of India as a Hindu country, and of Hinduism as an Indian religion.

These tendencies have continued, and since then, there has been a frequent eruption of communal violence. In 1992, Hindus incited to tear down the Babri mosque in Ayodhya, which they believe was intentionally and infuriatingly built over the site of Rama's birth. Tensions have intensified by attempts to convert Hindus to other religions.

Deities

The teaching of Hindu gods and goddesses are more complex than any other living religion.[54] Hinduism, unlike other religions, has no single founder, no single scripture, and no agreed system of teachings. The pantheon of gods and goddesses has gone through many changes over time. New deities have emerged, and some have been downgraded in importance; for this reason, more complexity came about in the Hindu religion.

The Vedic Aryans showed admiration for the controlling forces of nature. The natural phenomena were thought to be out of human control and brought misery to humans; therefore, the deities who presided over these phenomena conciliated to gain favor and protection. They praised through the hymns of the Rig-Veda. The three most significant deities are Brahma, who creates the universe, Shiva, who destroys the universe, and Vishnu, who preserves the universe. They are called trinity (Trimurti).

Brahma: In the Vedic period of ancient India known as Prajapati (Sanskrit: Lord of Creatures), Brahma's job was the creation of the world and all creatures. His name should not be confused with Brahman, who is the supreme God force present within all things.

Vishnu: A dark-blue-skinned entity with four arms, Vishnu is the preserver and protector of the universe. He is the most popular god of Hindus. His responsibility is to return to the earth in troubled times and restore the balance of good and evil. Occasionally he enters the world as Avatar. Hindus believe he will incarnate one last time close to the end of this world.

Shiva: Shiva takes many forms within the religion, with his main feature being the third eye on his forehead. He is often represented by a phallus-like lingam, a cylindrical pillar found in temples dedicated to Shiva. Hindus believe his powers of ruination and recreation are used to destroy the illusions and imperfections of this world, paving the way for beneficial change. According to Hindu belief, his demolition destruction is not arbitrary, but constructive. Therefore, he is seen as the source of good and evil.

Other major Hindu deities include Ganesh, Krishna, Indra, Rama, Lakshmi, Radha, Shakti, Durga, Varuna, and Saraswati.

Leaders (Priests)

Hinduism is not a centralized religion. Therefore, there is no comprehensive religious institution to appoint priests who learn scripture and sacraments at schools. Priests are generally trained from childhood by their parents and community leaders to become spiritual leaders. This training conveys knowledge about the rituals, mantras, chants, and unique hand gestures they will use later in their profession.

Priests are the extreme inciting factor as to why ordinary people of the Hindu religion became involved so profoundly in a sacred path—spiritual teachers and guides called gurus. Some gurus rise to local or national prominence due to their mastery in various aspects of religious knowledge and practical wisdom, and they gain widespread acclaim.

Hindu priesthood is a hereditary position, with families mostly tracing their priestly lineages back many generations. They often come from the Brahmin caste, whose primary function is education. Brahmin priests have retained their influence over the centuries. India's traditional caste system is in a fast-changing mode for adapting to the modern view on self-determination and spiritual purity, so some temples are expanding their ministry beyond the Brahmin class; as mentioned earlier, a guru can be non-Brahmin. Traditionally, only men were allowed into the priesthood, but now based on specific movements, a woman can be a priest.

A Hindu priest is called pandit (Sanskrit: an honorary for a learned man or scholar). A temple priest is called pujari (Sanskrit: worshipper), one who performs puja (Sanskrit: worship). Each temple will have its priest, or for a more significant temple, a whole team.

Hindu beliefs range from polytheistic to henotheistic, revering one God while recognizing the existence of many other gods. The roles taken by pujaris within their temples vary from community to community. Pujaris are mostly concerned with continuing the sanctity of their temples and devote much of their time to meditation, personal worship, and kindheartedness.

Scriptures

The primary source for the understanding of the historical meaning of Hinduism can be found in the sacred Vedas (Sanskrit: knowledge). The earliest holy texts possibly were introduced between 1200 and 1000 BC to Indians by the Aryans, who settled in northwest India around 1500 BC. Hindus believe that the script was received by sages directly from God and passed on from generation to generation by word of mouth. These scriptures can be amended to meet the changing times. The following are more information on these sacred scriptures.[55]

Shruti: Shruti texts are composed in the older form of Sanskrit called the Vedic, which describe the Vedic deities who were the controlling spirits of the forces of nature, and they are four Vedas: Rig-Veda, Sama-Veda, Yajur-Veda, and Atharva-Veda. Each Veda is divided into two sections, Samhita and Brahmana. Samhita contains hymns, and Brahmana explains those hymns and instructs how and when to use them. The Shruti Vedas cannot be altered or modified. Also, the Vedas have some highly philosophical discussions called Upanishads (Vedanta). Upanishad means to sit down beside their teachers (Sanskrit: *upa*—near, *ni-shad*—to sit or lie down). Upanishad has core teaching from mystics about Brahman (Highest God, Force Present within All Things), Atman (the soul or self), and Moksha (free from the cycle of death and rebirth). The most popular Upanishads are Aitareya, Brihadaranyaka, Chhandogya, Isha, Katha, Kena, Mandukya, Mundaka, Prashna Shvetashvatara, and Taittiriya.

Smriti: The Smriti scriptures have only secondary authority. All the scriptures other than Vedas fall under the Smriti category. The texts include relevant religious manuals. The passages relate to law and social conduct, like the Manu Smriti (Laws of Manu). The Puranas versions include Bhagavata, Skanda, Vayu, Markandeya, and Chandi. The epic texts are Ramayana, Mahabharata, and Bhagavad Gita. Some of Smriti text are as follows:

- *Manu Smriti* (Sanskrit: Laws of Manu), also called Manava-Dharma-Shastra, is composed of the legendary first man and lawgiver, Manu. Its composition likely dated to first

century CE. These texts instruct Hindus on rights, moral conduct, dietary restrictions, pollution, social duty, and the caste system.
- *Puranas* (Sanskrit: Ancient) is a collection of myths, genealogy, and stories about the gods, goddesses, sages, and heroes. It includes a detailed description of celebrations and festivals.
- *Ramayana* (Sanskrit: Rama's Journey) is an epic poem dating back to 300 BC, attributed to the poet Valmiki. It tells the tale of Rama and his journey to find and save Sita, his wife, after a demonic king abducted her. Rama eventually finds Sita with the help of a monkey army.
- *Mahabharata* (Sanskrit: Great Epic of the Bharata Dynasty) is one of the two ancient epic poems in human history. The other is the Ramayana. The poem is made up of almost 100,000 couplets and 1.8 million words. Its composition goes back to 900 BC, and it was not in writing for a couple of centuries. The poem is attributed to the sage Vyasa, and tells a story of the war between a family of brothers. The date and the occurrence of the war are much debated.
- *Bhagavad Gita* (Sanskrit: Song of God) is a 700-verse poem recorded in the Mahabharata. It is composed in the form of a conversation between Prince Arjuna and Krishna (a deity and avatar of the god Vishnu) before the battle. Written perhaps in the first or second century CE, it is commonly known as the Gita. They discuss the importance of selflessness, dharma (devotion), and the pursuit of moksha (liberation).

Darshana: Darshana, or Darshan (Sanskrit: viewing), are six different systems of philosophy. They are a religious, philosophical system because their foundation is in Vedic. Hindus observe a deity, revered person, or sacred object, and the experience is considered to be reciprocal and results in the human viewer receiving a blessing. The six Hindu philosophical systems of Darshan are Sankya, Purva Mimamsa, Uttara Mimamsa (Vedanta), Yoga, Nyaya, and Vaisheshika. Non-Hindu Darshanas include Buddhism and Jainism.

Tantras (Agama and Nigama): Besides the Vedic disciplines, Hinduism has another set of regulations, called the Tantras (Sanskrit: loom, the interweaving of traditions and teachings as threads). In real life, Tantra is one of the most prominent Indian traditions, representing the practical aspect of the Vedic tradition. In Tantra, God is looked upon as both a male and a female principle, called Shiva and Shakti. This view, taken to its extreme, holds that Shiva without his Shakti is like a corpse. They are always inseparable.

Denominations

Hinduism is comprised of diverse beliefs, with a hundred gods and goddesses. Hindus are free to worship one or a handful of gods. All Hindus are trying to obtain liberation; therefore, they participate in a varied set of beliefs and rituals. Around 500 CE, Hinduism became sectarian. As a result, most practicing Hindus belong to a denomination of Hinduism. There are many sects; among them are four significant denominations or sects: Shaivism, Shaktism, Vaishnavism, and Smartism.

Shaivism: Shaivism is the sect that is worshipping Shiva as the supreme reality. The Shaivas are self-disciplined, holy, and regard philosophy highly. They practice a type of yoga known as Bhakti-raja-Siddha, or yoga leading to oneness with Shiva within. They wear ash on their foreheads as a sign of their devotion, and temples devoted to Shiva are decorated with a lingam (Sanskrit: shaft of light; it is a sign for Shiva as presented by a phallus). Shiva is the consort of Shakti.

Shaktism: Shaktism worships the Divine Mother Great Goddess Shakti and her many other forms. Some of the types are fierce as a rumble of thunder, and some are gentle as a breeze. She is typically called Devi. Shaktism is closely related to Shaivism due to traditional beliefs that Shiva without his Shakti is like a corpse. They use magic, fire walking, chants, yoga, tantric meditation, and a variety of rituals to manipulate cosmic forces through the human body.

Smartism: Smartas are a nontraditional and progressive movement. They worship all forms of Supreme Being: Vishnu, Shakti, Shiva, Ganesh, Skanda, and Surya. They are roughly nonsectarian, and they follow a philosophical meditative path that emphasizes the oneness of man with God through understanding. Adi Shankara, who lived from 788 to 820 CE, was one of the most revered philosophers and theologians who gave Hinduism the new liberal denomination called Smartism. The Puranas, Bhagavad Gita, Ramayana, and Mahabharata are relevant scriptures in this nonsectarian tradition.

Vaishnavism: Vaishnavites worship and adulate Vishnu as the Supreme Being, along with Rama and Krishna and his many avatars and incarnations. Vaishnavites are profoundly devotional, practice meditation, and their religion has many saints, temples, and scriptures. Their scriptures draw heavily from Bhagavad Gita and Upanishads.

Beliefs

Hinduism, unlike other religions, has no single founder, no single scripture, and no agreed system of teaching. Hindus have no simple set of rules to follow. Hinduism is a religion, but it is mostly a broad way of life. Hinduism is one of the oldest religions, having an extensive spectrum of practices and beliefs, some of them easily tied to primitive pantheism. Some basic principles identify the Hindu belief and practices.[56]

The Purusarthas, or four goals of life: Purushartha (Sanskrit: object of human pursuit or aims of the human) belief is that human beings need to seek all four goals, although people may have unique talents in one of the four.

- *Artha* (Sanskrit: wealth or property) is the first goal of life. It means material prosperity through constructive work and achieving worldly well-being. It is closely linked to government activities, which maintain the general social order.
- *Kama* (Sanskrit: love, desire, and pleasure) is the second goal of life: to have fun. In Indian mythology, Kama is the counterpart of Cupid. He is the Hindu god of love. The Kama is regarded as one facet of a well-rounded spiritual life. Another scripture, Kama Sutra, was compiled by

Hindu philosopher Vatsyayana Mallanaga in the third century CE. The title of this script is derived from the Sanskrit Kama, which denotes yearning and desire, often with a sexual undercurrent (and referring to the Hindu god of love, one-fifth of this Scripture is devoted to sexual positions).

- *Dharma* (Sanskrit: religious and moral law, ethical) is the third goal of life. It refers to a duty-driven way of living in cooperation with your fellow human beings. In general, it refers to one's class and status in life. In Hindu culture, your dharma will be determined by the family and clan you were born into. If you are born into the family of a carpenter, your dharma or career will be as a carpenter.
- *Moksha*, also called Mukti, from the root of much (Sanskrit: to free, release, let go, even enlightenment). This is the fourth goal of life, the pursuit of spiritual liberation, salvation, and freedom from samsara. Samsara, the cycle of death and rebirth, is bound to live in the material world.

Karma and Rebirth: Karma (Sanskrit: action, word or deed, also spiritually cause and effect) came from Hindu tradition and philosophy. It states the act itself will hold its originator responsible and accountable; in other words, the law of karma says that a human reaps what they sow. The person with bad karma spiritually or ethically could suffer being reborn many times in lower castes of humans, or even worse, in an unfortunate animal. The bottom line is what you do today determines what you will be after rebirth. A person with good karmas at every rebirth moves up to a better class, and can even be a deity. A person with bad karma, or sinners, may spend a while in hell, which is called Naraka (Sanskrit: underworld), ruled by Yama (Sanskrit: moral discipline), the god of death, and is located south of the universe.

Samsara and Moksha or Moksa: Hindus believe Samsara (Sanskrit: action, the process of reincarnation [transmigration]) is the whole process of rebirth or reincarnation. The life's ultimate goal is Moksa, or nirvana (Sanskrit: liberation), which is the realization of your relationship with God and being detached from all worldly concerns.

The Soul: In Hinduism, soul is known as Atman (Sanskrit: eternal self, essence, soul, spirit). Hindus believe Atman or soul indicates our true self or nature, which underlies our existence separate from body and mind, beyond ego, and includes the belief that all creatures have a soul. This soul is different from universal soul, known as Brahman. Many schools of Hinduism identify Atman as indistinct from a global soul.

Afterlife: In Hindu culture, a cremation should take place on the day of death. Usually, they desire to die at home, and the head of the household of the deceased will light a candle. The body will be placed in the entranceway of the house with the head facing to the south, which is ruled by Yama, the god of death. The body is washed, consecrated with sandalwood, and wrapped in cloth. The corpse is then carried to the funeral pyre by the male relatives. Prayers are said to Yama. Sometimes the name of God (Ram) is chanted. On the funeral pyre, the head of the corpse is positioned to the north toward the Kubera (Sanskrit: god of wealth) and the feet toward the south, to the realm of Yama. Usually, the chief mourner sets fire to the pyre, and afterward, the cremation ashes will be thrown in the river Ganges, which is considered the most sacred water.

Four Stages of Life and Their Rituals

In Hinduism, human life is believed to consist of four stages. These stages are called ashramas (Sanskrit: toll, fatigue, making an effort toward liberation), and everybody should ideally go through these stages.

- Brahmacharya, the student stage, or in Sanskrit this is composed of two words: Brahma (God, creation) and charya (to follow).
- Grihastha, or householder stage.
- Vanaprastha, or hermit stage. In Sanskrit it is composed of two words: vana (forest) and prastha (going to), which can translate as "retiring to the forest."
- Sannyasa, the wandering ascetic stage. In Sanskrit it means "purification of everything."

Caste System (Discrimination)

The caste system divides Hindus into very rigid hierarchical groups based on their duties. Manu Smriti (Sanskrit: Laws of Manu) goes back to 1250–1000 BC and part of creation. This law comes from the Rig Veda. The hymn of Purusha[57] explains that the universe was created out of the parts of the body of a single cosmic man, Purusha, when his body was offered at the primordial sacrifice. The four classes came from his body: the priestly class (Brahmans) from his mouth, the warrior class (Kshatriya) from his arms, the peasant class (Vaishya) from his thighs, and the servant class (Shudra) from his feet. The members of non-Hindu groups were considered Dalits (Sanskrit: broken, scattered, outcasts, or lowest class).

After almost three thousand years of the caste system (discrimination), in 1950, the government of India attempted to acknowledge and delegitimize the caste system as the basis of order and consistency of society to rectify historical injustices and supply a level playing field to the traditionally disadvantaged lowest class to provide jobs and education. In 1989, the government made another modification to quotas to extend and include another group called the OBCs (Other Backward Classes), which fall between the established upper castes and the lowest.

Rituals and Customs

Hindus are trying to get to oneness by performing the practices mentioned above. Some underlying methods are standard through its many interpretations. We will define some of the most fundamental of these practices, and include rituals, worship, meditation, recitation, family-oriented rites of passages, pilgrimages, and annual festivals.

Purification and repentance: Purification and repentance are two of the most crucial parts of the practice of rituals and ceremonies. Positive punya (Sanskrit: deeds, virtue, purity) is necessary to counteract papa (Sanskrit: irreligious act, immoral act, evil act, vicious act), like charity and performing daily rites. Hindus believe that many human activities detach the person from the divine. Before the beginning of any ceremony, the person who performs the service will go through the process of purification, which includes bathing and avoiding eating meat, working with dead things, and taking the life of an animal.

Worship: Hindu worship is called puja (Sanskrit: reverence, adoration), which can take place at home or in a temple. The majority of the devotions take place in the house because Hinduism is part of daily life. Almost every home has a household shrine with a statue representing a god called Murti (Sanskrit: the embodiment of the divine, the Ultimate Reality). Hindus burn incense, praise, chant mantras, wash Murti, meditate, and offer food sacrifices. At temples, the pujaris (Sanskrit: worshipper) is a member of the Brahman priestly class who performs puja that is identical to home worship.

Samskara[58] *and pregnancy*: During pregnancy, a few rituals are practiced, like the third-month ceremony called Pumsavana,[59] which is intended for the protection of the fetus. In the seventh month is a service called Simantonnayana,[60] designed for the health of the mother, and prayers are offered for the child to ensure healthy mental development. Pumsavana and Simantonnayana are rarely performed at present, and taking their place is *dohale jevan*, which takes place in the seventh month of pregnancy. It involves gathering women with food, gifts, flower, fruits, and singing songs.

Birth: At birth, the most significant event is called Jatakarma[61] (Sanskrit: newborn infant), a welcoming of the child into the world. It is a private one performed by the parents. The father puts honey in the baby's mouth, rubs ghee on its lips, and recites the name of God in the baby's ear.

Marriage: A Hindu wedding incorporates many timeless rituals and customs. In the past, these traditions and rituals would extend over several days, but in today's society, all the ceremonies are completed the day before and the day after. Marriage in India is an arranged marriage. When a suitable spouse has been found for the daughter or son by the parents, their approval is a must. The dates of the wedding are decided by the parents in consultation with an astrologer to pick a lucky time for the ceremony and, most importantly, to agree on the dowry, which the bride's family pays to the groom's family.

Bride and groom follow a ritual of offering their right hands, which are symbolically bound together with cotton thread that has henna for coloring their hands, in front of the fire, a holy symbol in Hindu religion. The bride and groom then recite seven vows to each other when circling the fire. It is called saptapadi[62] (Sanskrit: seven steps), and it is a central part of a Hindu marriage ceremony.

Death: For death, ritual, and ceremony, see the section "Afterlife."

The Sacred Thread Ceremony: The Upanayana (Sanskrit: the act of leading to or near) ceremony is restricted to three upper classes: Brahmans (priests and teachers), Kshatriya (warriors and rulers), and Vaishya (merchants and tradesmen), and is performed when five to twenty-four years old. The participant is given three strands of sacred thread, which represent three promises they make: to respect expertise, their parents, and the society.

Ceremonies and Festivals

Hinduism has a multitude of festivals during the year. The following are some popular and more significant festivals.

Diwali: Diwali or Deepavali is the festival of lights, the most famous festival on the Indian subcontinent. The underlying essence of Diwali revolves around fire superseding darkness or, figuratively, the triumph of good over evil. They celebrate the return from exile of Ram and Sita.

Holi: One of the major festivals of India, Holi[63] is celebrated with enthusiasm on the full moon day in March. It is a celebration of the slaying of the demoness Holika by Prahlada, a devotee of the deity Vishnu.

Rama Navami: Rama Navami celebrates the birth of Lord Rama, son of King Dasharatha of Ayodhya.[64] Rama was an incarnation of Vishnu and the hero of the Ramayana. The ritual consists of songs, dances, and puja dedicated to Rama.

Krishna Janmashtami: Krishna Janmashtami is the celebration of the birth of Lord Shri Krishna, the incarnation of Lord Vishnu.

The faithful celebrate with dance, song, and puja dedicated to Lord Shri Krishna.

Vijayadashami: Also known as Dussehra, this festival marks the triumph of Rama, the incarnation of Vishnu, over the ten-headed demon Ravana,[65] who abducted Rama's wife, Sita. This festival celebrates the beginning of harvest season, and it includes a ritual to fertile soil and fertility.

Navratri: Navratri is also called Durga Puja. This festival is held over nine days during September and October in honor of the divine feminine. It symbolizes the triumph of good over evil. Navratri annihilated the demon Mahishasura[66] after a relentless battle lasting nine days and nights. Navaratri is a festival in which God is adored as Mother.

Yoga

Vedas, which are a collection of hymns received by sages as Shruti, or divine revelation, are considered the most holy texts of Hinduism. Amazingly, within these texts, the foundations of yoga (Sanskrit: a group of mental, spiritual, and physical practices, or yoke; the connecting link) are established. Hinduism offers many different paths to reach God. There are four primary paths:

1. Bhakti Yoga, the way of devotion
2. Jnana Yoga, the path of rational inquiry
3. Raja Yoga, the path of mental concentration
4. Karma Yoga, the path of right action

Summary

Hinduism is one of the oldest world's religions of the Indian subcontinent. Hinduism, with significant roots from the primitive faith, became the advanced religion of Hinduism. (Religion evolved based on fear-emotion in primitive religion, and we can see the root of primitive religion now in Hinduism.) They revere snakes, bear, tiger, the mouse, cows, peacocks, monkeys, the Ganga and Jamuna Rivers, and much more.

Fear is the best tool to keep people under tight control by leaders and religious authorities. In Hinduism, there are two primary components.

1. *Karma and rebirth*: Karma (Sanskrit: action, word or deed, also spiritually cause and effect). The law of karma (proposed by religious leaders) says that a human reaps what they sow. The person with bad karma mentally or ethically could suffer being reborn many times in lower castes of humans or, even worse, in an unfortunate animal.
2. *Samsara and Moksha*: Samsara (Sanskrit: action, the process of reincarnation [transmigration]) is the whole process of rebirth. Moksha (Sanskrit: liberation) is the realization of your relationship with God and being detached from all worldly concerns.

CHAPTER 5

Zoroastrianism

View to the Zoroastrian temples ruins and the Tower of Silence, Yazd, Iran (Copyright iStock by Getty Images/credit: Nmessana and Fidan Babayva).

Zoroastrianism

Zoroastrianism is believed to be the first world's oldest monotheistic religion, and its founder was a prophet named Zoroaster (Persian: Zarathushtra).[67] Zoroaster's date of birth, death, and the first place where he preached isn't precisely known and is a matter of dispute among scholars. The location where his family was living has been established. He was living with his family along the Oxus River (the Amu Darya, along the northern border of Afghanistan). During his adult life, he moved to Bactria or Bactriana (present province of Balkh in Afghanistan). Tradition suggests Zoroaster was born in 628 BC and died in 550 BC. Based on archaeological evidence and linguistic comparison, modern scholars believe that Zoroaster must have lived sometime between 1500 and 1200 BC. Some classical Greek historians, like Diogenes Laertius, place Zoroaster's life at 6480 BC.[68]

It is believed the ancient Aryans were living from north of Hindu Kush to the Oxus River and beyond. This area was part of greater Khorasan, also spelled Khurasan, and consisted of the eastern part of Iran, Afghanistan, Tajikistan, Turkmenistan, and Uzbekistan. Khorasan was also recognized as Ariana, or the land of Aryans. Zoroastrianism was the dominant religion in Khorasan.

The first Persian Empire, founded by Cyrus the Great at around 550 BC, became one of the largest empires in history, stretching from Europe's Balkan Peninsula in the west to India's Indus Valley in the east. At this time, Khorasan became part of the Persian Empire, and the official religion of the Persian Empire became Zoroastrianism.

The origin of the Vedas scripture of Hinduism may be between 1800 and 1500 BC, when the Aryans (Indo-European, Indo-Aryan) invaded India. According to some scholars,[69] the Aryan language gained ascendency over the descriptive words on the Indian subcontinent. Based on this assumption, both the Sanskrit language and the Vedic religion foundational to Hinduism are attributable to Aryan culture.

Zoroaster arose at a time when the polytheism system was the religion of Egypt, Mesopotamia, and India. He was practicing as a priest in those pantheons of the gods. At the age of thirty, while he was attending a spring festival of pagans at the bank of a river, he was struck by a blinding white light, and a Shining Being revealed himself as Vohu Manah (Good Mind). Vohu Manah led Zoroaster to Ahura Mazda (the Wise God) and five other radiant beings, which are called the Amesha Spentas (Holy Immortals). Ahura Mazda called Zoroaster the first and last prophet.

During this meetup, Zoroaster asked many questions, and the answers given to him are the foundations of the Zoroastrian religion.

Also, Vohu Manah told him about hostile spirits, or Angra Mainyu, and Asha (truth), and Druj (lie). With this revelation, Zoroaster was enlightened; he decided to continue telling people about Asha for the rest of his life.

Zoroastrianism was one of the most prestigious religions in the world during the rise of the Persian Empire (600 BC–600 CE). During this period, in 330 BC, Alexander the Great invaded Persia, burned the Avestan (Manuscript of Zoroastrianism, saved at the Royal Library at Ishtakhr). He killed priests and many other upholders of Zoroastrianism.

During the rise of Islam, the Persian Empire under Sasanians faced turmoil amid military, economic, political, and social weakness after the execution of King Khosrow II in 628 CE. The Islamic Empire took advantage of the Persian weaknesses during the reign of King Yazdegerd III (Persian: Yazdiger).

The Islamic Empire, under the reign of Caliph Umar ibn al-Khattab, mobilized a strong military mission. This mission started in 636 CE, and they invaded Persia in 651 CE after the execution of King Yazdegerd III. The Arab invaders killed thousands of innocent people, including children and women. They burned temples, burned and destroyed all the scriptures, yet with all these ferocities and massacres, they could not eradicate Zoroastrianism.

Finally, they used a very divested strategy by allowing the Persians to keep their religion, but they must pay a considerable tax. They cut the lifeline by imposing huge tax on the family, for each of the children and women. Soon after this policy, many Persians fled their country to Surat, and later they settled in Bombay, where they remain.

Zoroastrianism is the oldest religion, and founded monotheistic religion. All Abrahamic faiths, such as Judaism, Christianism, and Islam, borrowed many rituals from Zoroastrianism, which will be discussed in later sections.

Deities

Ahura Mazda: Ahura Mazda (also known as Ahuramazda, Harzoo, Hurmuz, Ohrmzad, "Wise Lord," "Spirit") is the Supreme God of the Zoroastrian faith. He is the creator of the universe, with all good things in it. He is just, caring, omnipotent, omnipresent, omniscient, the source of all happiness and goodness. He is changeless, moving all things while not being moved by anyone. Ahura Mazda, along with the Messiah or Saoshyant and the Amesha Spentas (Holy Immortals), will defeat the Angra Mainyu to make the world free from evil.

Angra Mainyu: Angra Mainyu (Avestan: destructive spirit) refers to evil in the Zoroastrian religion and dwells in a hellish abyss. He has opposed Ahura Mazda since the creation of the universe. He destroyed the first man, Gayōmart, who was created by Ahura Mazda, with disease. Angra Mainyu is the originator of death and all that is evil in the world.

Spenta Mainyu: Spenta Mainyu is the Holy Spirit of Ahura Mazda. Zoroastrians believe that the Spenta Mainyu protects and maintains the order in the universe.

Spenta Amesha: Spenta Amesha is a group of six holy immortals that were created from Ahura Mazda's soul to guard his creations against the Angra Mainyu. The newly created immortals are

1. Khashathra, the Righteous Power, who became the protector of the sky,
2. Haurvatat, the Peace and Perfection, who became the protector of water,
3. Spenta Armaiti, the Holy Devotion, who became the protector of the earth,
4. Ameretat, the Immortal, who became the protector of plants,
5. Vohu Manah, the Good Mind, who became the protector of animals, and
6. Asha Vaishta, the Justice, who became the protector of fire.

These six immortals, three females and three males, are often pitted against daevas[70] (demons of the desert).

Saoshyant: Saoshyant, or the final Messiah (Avestan: one who brings benefit), is said to arrive on the earth before the end of time. With six helpers, Saoshyant will lead the world to raise the dead to life and cleanse all of them at the molten metal rivers, including those damned. He will prepare for them white Haoma, the ritual drink of Zoroastrians. They will be led to the new world or Frashokereti, the end of the present state of the world, and live in peace with Ahura Mazda.

Vizaresh and Daena: Vizaresh and Daena are the divines who are present at the Chinvat Bridge to guide the departing souls. Those who were pure and righteous will be accompanied by Daena to the House of Songs, and those who were wicked will be guided by Vizaresh to the House of Lies.

The Trinity Judges: The trinity judges are divinities who judge the soul's worthiness to cross the Chinvat Bridge to enter the House of Songs or to be banished to the House of Lie. The trinity judges are Rashnu (who represents justice), Sraosha (who represents obedience), and Mithra (who represents truth).

Yazatas and daevas: Yazatas is the Avestan word for worthy, worship, or holy, and is an order of angel created by Ahura Mazda, just like Amesha Spenta and Daena. He helps Ahura Mazda maintain the world order and fights back the forces of the demon Angra Mainyu. Daevas are the counterpart of Yazatas. The Gathas, the oldest part of Avesta, mentions three daevas: Druj, Aka Manah, and Aeshma. Angra Mainyu created them; they are disrupting the world order and distracting humans from good deeds.

Gayōmart: Gayōmart was the first human created by Ahura Mazda and the first to worship him. Gayōmart was attacked by Angra Mainyu, who gave him sickness and disease, but he didn't realize that before his death, his semen fell to the ground. From that semen grew a rhubarb plant. After many years, a man (Mashya) and a woman (Mashyana) grew out of the plant. The couple promised the Wise God that their children would fight with him in his battle against Angra Mainyu (the Evil Spirit).

Leaders

Zoroaster was a leader who created the new idea of one God, while the neighboring societies of the Indian subcontinent and Middle East believed in hundreds of gods and had a caste system. Zoroaster was a leader who never used swords and killing to impose his ideas and beliefs on others. Zoroastrianism was the largest religion in the world until Muslims eliminated them.

Zoroaster: Zoroaster (Zarathushtra) is one of the most exceptional leaders, reformers, and prophets of his time. His theology and thinking influenced many other religions, like Judaism, Christianity, Islam, and Buddhism. At this point, it isn't known if Zoroaster is the prophet's real name or just a title, like Buddha or Christ.

Zoroaster is known to us primarily from the most crucial portion of the Yasna texts, which are the five Gathas. From the Gathas are seventeen great hymns composed by Zoroaster himself. He was born into a very affluent family from the Spitama clan; his father's name was Pourushaspa (with gray horses). His mother's name was Dughdhova (who has milked cows).

He started at the age of seven in a school of the priesthood similar to the ancient Vedas of India, which he completed by the age of fifteen. During his school years, he showed signs of intelligence and wisdom. At the age of twenty, he left home and traveled to a different part of the country and spent a lot of time in meditation. At the age of thirty, Zoroaster received the revelation. (See the previous section for details.)

This period of ten years was possibly the best time to think and plan a new idea, how to succeed, and influence those who believed in so many gods. The polytheism religion was ancient, and did not answer people's demands. However, it is not easy to convince people to accept new ideas.

Zoroaster's idea of one God did not go very well during his first ten years of his activity and preaching. He was only able to convert his cousin Maidhyoimanha.

The local religious authorities opposed Zoroaster's new ideas because he called their gods demons and evil spirits, and Zoroaster was against their rituals and especially sacrifices.

Finally, after all those years of defeat, he realized he needed to couple his preaching with force and authority. He found the perfect place: the palace of King Vishtaspa in Bactria (Balkh).

The king and the Queen Hutaosa had heard of religious debate between Zoroaster and the religious leaders of the kingdom. First, he jailed him, but later, his favorite horse was sick, and the king asked Zoroaster to cure the horse.

Zoroaster grew up around animals, also took care of unhealthy horses, and he had a love for animals. Zoroaster's father, Pourushaspa (with gray horses), was a horse breeder. The horse was cured, and King Vishtaspa decided to accept Zoroaster's ideas and made Zoroastrianism the official religion of the kingdom.

Zoroaster continued preaching until his death at the age of seventy. During his lifetime, he did not have much success in spreading Zoroastrianism in the vast land of Khurasan. After centuries, Zoroastrianism was spread beyond Khurasan and Persia through military invasion and imposing this new religion on the vanquished. During the reign of Cyrus the Great,[71] Zoroastrianism was the official religion of the largest empire in the world. Cyrus invaded Mesopotamia and ended the captivity of the Jews from the Babylonians and let the Jews go to Jerusalem, and rebuilt the temple in Jerusalem. In the Bible, he is the liberator of the Jews who were captive in Babylonia.

Scriptures

Avesta: The Zoroastrian scripture is called Avesta, and was written in the Avestan language. There is no known date when the Avesta was published. Since it is believed Zoroaster lived in Bactria, it is a possibility that the Avesta was published in Bactria. The original Avesta is said to have consisted of twenty-one Nasks, or volumes. In 330 BC, Alexander the Great invaded Persia. As a result of the invasion, manuscripts were lost, dislocated, or burned, and priests and many other upholders of Zoroastrianism were killed. The remaining manuscripts of Zoroastrianism are saved at the Royal Library at Ishtakhr.

Zoroastrianism

The Avesta contains the teaching of Zoroaster: law, liturgy, and cosmology. The remaining Avesta is now a fragment of the original scripture. A great deal of it, some say two-thirds of it, was lost or destroyed when Alexander the Great invaded Persia.[72] The original is said to have consisted of twenty-one books or volumes. The most sacred part of the Avesta is from the Gathas, the seventeen great hymns Zoroaster composed himself.

Zend-Avesta is the commentary to the Avesta.

The first compilation of the Avesta was assembled from the remnants of the original Avesta during the Sassanid Empire (224–651 CE). The Avesta is organized in five parts.

1. Yasna means service, dedication, and prayer. The liturgical part contains seventeen songs or hymns, known as the Gathas, which scholars believe Zoroaster himself wrote. Priests recite the Yasna when performing priestly duties. The Yasna also contains the rite of sacrifice and preparation of Haoma.[73]
2. The Visperad is always recited with the Yasna. The Visperad is recited in seasonal gathering feasts and Nowruz (New Year's Day). It also contains reference to several Zoroastrian spiritual leaders.
3. The Vendidad, or Videvdat, contains Zoroastrian rituals against daevas (desert demons) and evil; it also includes the story of creation and the first man.
4. The Yashts includes twenty-one hymns and songs dedicated to various angels, Zoroastrian ideas, and ancient heroes.
5. The Khurda Avesta, or Little Avesta, is a collection of minor hymns, text, and prayer for special occasions.

Bundahishn: Bundahishn[74] is another Zoroastrian text providing an account of the creation myth, the origin of man, and the nature of the universe. Written in Pahlavi, it dates to the eighth or ninth century CE.

Denominations

Zoroastrianism was the second largest world religion without divisions and denominations and a religion of one founder. During the rise of Islam, the Islamic Empire took advantage of Persian weaknesses during the reign of King Yazdegerd III and mobilized a strong military mission. The Arab invaders killed thousands of innocent people. Some Zoroastrians fled their country and settled in India, with a small population of Zoroastrians who survived in Iran and elsewhere. They divided into three parts: Parsi, Irani, and Gabar. Presently, the total population of the three Zoroastrian divisions is approximately 200,000 followers.

Beliefs

Zoroastrians believe in one God, who is Ahura Mazda (Wise God). He gave humans free will and made each of us morally responsible for our actions.

Creation

The creator of the universe is Ahura Mazda (the Wise Lord), according to the Book of Creation (the Bundahishn, which dated later than Avesta, around the sixth century CE). In the beginning, there was nothing other than Ahura Mazda (the Wise God), who dwelled in the endless light and goodness above, and Angra Mainyu (the Evil Spirit), who resided in the darkness and ignorance below. Between them lay only space.

Ahura Mazda decided to make different creations. He made sky out of metal, shining and bright; pure water; earth round and flat, with no mountain and valleys; plants, moist and sweet, without bark and thorns; and animals big and small. Finally, he created the first man from a clay, whom he called Gayōmart.[75] He was tall and handsome. Lastly, fire was distributed among the whole of creation, as noted in the story below.[76]

The Angra Mainyu (the Evil Spirit) looked out of his darkness to see Ahura Mazda's fabulous creations. Ahura Mazda called on him, and said, "Angra Mainyu! Help my creature and praise them so that you will be immortal." Angra Mainyu responded, "Why should I help and admire your animals? I am powerful! I will destroy you and your creature." He went back to his Darkness and created witches, monsters, and demons to attack the Endless Light. Ahura Mazda was all-knowing, so he created six spirits (Amesha Spentas), the Holy Immortals, from his soul, to guard his creations against the Angra Mainyu.

So, the Angra Mainyu (the Evil Spirit) attacked the Lord's creations one by one. He did not have success, except, on the attack on the first man, Gayōmart. The Evil Spirit gave him sickness and death. Angra Mainyu was very happy that he became victorious against the Ahura Mazda, but he didn't realize Gayōmart's semen fell to the ground before his death. From that, the semen grew a rhubarb plant. After many years, a man, Mashya, and a woman, Mashyana, grew out. The couple promised the Wise God that their children would battle against the Angra Mainyu (the Evil Spirit).

Soul: The soul, or Urvan or Ravan, is immortal, and every creature has one. The soul exists before birth with its guardian angel; after delivery, it combines with the physical body on the spirit a person chooses, and it's within the person's control with their free will. The fate of the human mind depends on thought, words, and deeds. The soul will get an afterlife, based on what the person did in this life. On the third night after death, the soul leaves this world and enters the spiritual world. The soul stands before the Chinvat Bridge, or the Bridge of Judgment, where the person's deeds are reviewed. A kind soul is led to paradise, and an evil soul is directed to the darkness. At the end of time, all souls will be cleansed and return to the Wise Lord.

Ethics: Ethics and morals have a special place in Zoroastrianism. Walking on the Asha path, maintaining a good state of mind, being sincere and honest, having good thoughts and words, and doing good deeds, will make the Angra Mainyu weak. As a result, Ahura Mazda will defeat the Evil Spirit.

Fire: Fire is sacred in Hindu and Zoroastrian religions. Fire is the provider of energy, heat, light, and cooking. The early Aryans worshipped their gods with burnt offerings. Hindus didn't need fire to heat their dwelling for survival, but Zoroastrians were very dependent on fire because of the cold climate from the Hindu Kush to the Oxus River. Very likely, Zoroastrians influenced the Hindus to use fire in their ceremonies. The fire became a symbol of truth and purity, creating a path to Ahura Mazda.

Asha and Druj mean truth and lie. A path to become close to the Wise God is the basis of Zoroastrian belief and the counterpart of the Druj lie. Dualism may be seen in Zoroastrianism, but on the teaching of Zoroaster, Asha will be victorious in the end, and therefore dualism does not fit in Zoroastrianism. These opposing forces were borrowed from Zoroastrianism with a little modification in all Abrahamic religions: Islam in the Koran, Judaism in the Talmud, and other scriptures in the Bible.

Death: Zoroastrianism preaches that death is the work of the evil Angra Mainyu. After death, the soul, or Urvan, lingers for three days before departing to the kingdom of the dead. This place is ruled by Yima (Sanskrit: Yama), who had been the first king on the earth and the first man to die. To make it easy for the departing soul, the family would pray, make a blood sacrifice, and fast. The dead body is viewed uncleaned, and the family members, in even numbers, thoroughly wash the body of the deceased and dress the body in the recently washed old white cotton cloth. A prayer cap is placed on the deceased's head. Since the corpse is decaying matter, it must not be laid on the earth, fire, water, and air, which is considered blasphemy. Traditionally, the corpse was laid out on a *dakhma*, or "tower of silence." Nowadays, it would be cremation.

Chinvat Bridge: After the soul departs from the mortal body in three days, on the fourth day, the soul will cross the Chinvat Bridge, or "bridge of separator" (in Avestan it is called Chinvato Peretu). The soul crossing will be overseen by a trinity of judges: Rashnu, Sraosha, and Mithra. The souls who followed the Druj (lie and falsehood) will fall off the bridge as the bridge turns like a razor's edge, narrow and sharp, and they will be guided by Vizaresh to a hellish abyss, where they reside to the end of time. The soul who followed the Asha path, maintaining a good state of mind, will find the bridge wide and will be accompanied by the beautiful feminine, called Daena, and she will guide the soul to the House of Song or Garo Demana to reside there to the end of time.

Final Judgment: Zoroastrianism believes in the final judgment (Avestan: Frashokereti, making wonderful) at the end of time by the arrival of a last Messiah or Saoshyant (Avestan: one who brings benefit). All souls are resurrected and reunited with their resurrected earthly bodies, mountains turn to rivers of molten metals, and everybody will be immersed in the molten river. The pure and righteous will feel it soothing, like a warm bath, and those wicked ones who practiced evil will endure horrific suffering.

The molten rivers flow into the House of Lies to destroy the Angra Mainyu. After this, all humanity becomes immortal and lives in peace, exalting Ahura Mazda.

Zoroastrians were at the forefront in the idea of final judgment; afterward, the Abrahamic religions of Judaism, Christianity, and Islam borrowed this idea and modified it to fit their Holy Scriptures.

Sacrifice: Zoroastrianism is very close to Taoism when it comes to taking care of the environment. Zoroaster believed in personal responsibility. Animal sacrifice does not clear out a person's sin, obligation, or guilt, and therefore, he did not allow the sacrifice of an animal. Zoroastrians do not pollute the waters, rivers, land, or atmosphere, because everything Ahura Mazda created is pure. This has caused some to call Zoroastrianism the first ecological religion.[77] Also, Zoroastrianism has respect for the animals. like the dog, which is regarded as a clean and righteous creature that must be taken care of and fed. A dog's gaze is considered to be purifying and drives off demons.

Festivals and Holy Days

Holidays and festivals are the most critical part of the Zoroastrian faith, and are closely linked with seasons. The early Zoroastrians had presumably continued to use the pagan calendar of 360 days. It is unclear how it kept meeting the seasons. The Zoroastrian calendar has been adjusted many times, creating controversy among the Zoroastrian communities. Currently, three calendars are in use: Seasonal or Fasli, Shenshai or Shahanshahi, and Kadmi or Qadimi. The newer Fasli calendar incorporates the leap year, and the start of years is always on March 21. Therefore, the festivals are according to the Fasli calendar.

Naw-Ruz, also spelled Nowruz (Persian: the first day of the year), is the Zoroastrian New Year's Day, celebrated on March 21, on the day of the spring equinox. It is one of the two festivals mentioned by Zoroaster in the Avesta.

Pateti is the Zoroastrian New Year's Eve for Parsis. Many Parsis call Naw-Ruz Pateti. Pateti means "in Persian repentance." Parsis are eliminating the day of repentance and changing the meaning of Pateti to New Year's Day.

Khordad Sal is celebrated six days after Naw-Ruz as the birthday of prophet Zoroaster, and it is known as Greater Naw-Ruz. Zoroaster's birthday is unknown, and the day is selected symbolically. This festival is one of the significant festivals in the Zoroastrian calendar. Believers gather in fire temples for prayer and feasting.

Zarathust No-Diso, or Zarthost No Deeso, is a day of remembrance of Prophet Zoroaster's death. In the seasonal calendar, it falls on December 26. It is a festival of remembrance, with discussion and lectures on the life of Prophet Zoroaster in fire temples.

Yalda, or Shab-e Yalda, is one of the oldest festivals; it predates Zoroastrians. Zoroastrians celebrate the winter solstice as the triumph of good over evil, as nights get shorter and days get longer. Its celebration was borrowed by Islam.

Gahanbars (Gahambars or Ghambar), are six obligatory religious celebrations in Zoroastrianism. The very beginning of this festival dates back to the pre-Zoroastrian time. It appears that Zoroaster was very close to the agricultural revolution, and the agricultural people are accustomed to seasonal changes and celebrate survival for the next year. These days are

1. Maidozarem Gahanbar, a midspring festival in late April or early May;

2. Maidyoshahem Gahanbar, a midsummer festival in late June or early July;
3. Paitishahem Gahanbar, a harvest festival in September;
4. Ayathrem Gahanbar, bringing home the herds festival in October;
5. Maidyarem Gahanbar, a midyear celebration, originally celebrated on the winter solstice, now around the secular New Year; and
6. Hamaspathmaidyem Gahanbar, a ten-day festival at year's end in honor of the spirits of the dead.

Rituals and Customs

Given the fact that Zoroastrian history goes back over 3,500 years, they do not have many rituals regarding worship, in contrast with other Western religions. They focus on the ethics of Good Words, Good Thoughts, and Good Deeds. Zoroastrian rituals are cleanliness rituals, daily prayers, and are worshipped at home and in the fire temple.

Navjote: Navjote, or Sedreh-Pushi (Persian: wearing outfit) is a ceremony all young children between the ages seven and twelve must go through to receive their Sudreh, an undergarment to remind them of their modesty and to be trustworthy. They perform a Kusti ritual, a string made of seventy-two lines, one for each division of the Yasna, which they dress in for the rest of their life.

Purification: The purification ritual consists of three types:

1. Padyab, or ablution, is only washing the body or body parts.
2. The Nahn, or bath.
3. The Bareshnum, a complicated ritual, requires a dog whose left ear is touched by the candidate and whose gaze puts the evil spirit in flight; this ritual lasts a couple of days.

Zoroastrians must purify themselves before any gathering for rituals and worship.

Marriage: The Zoroastrian wedding ritual can last between three days and a week. The first thing that occurs is the signing of the marriage contract between the bride, groom, and parents, in the presence of a priest. Close relatives hold a white scarf over the new couple's heads while two crystallized sugar cones are rubbed together to sweeten the couple's life together.

Nirang-I Kushti: The Nirang-I Kushti, or girdle formula, is a rite of cleanliness that both males and females perform in preparation for prayer throughout the day. The kushti, or sacred thread-girdle, is a string long enough to pass three times very loosely around the waist, to be tied twice in a double knot, and the ends are left hanging behind. They raise the kushti to their forehead upon naming Ahura Mazda and tug it violently while naming Angra Mainyu.

Haoma/Yasna: The sacrifice of Haoma, which means "sacred liquor," is the production of the drink from the plant and drinking it. It is celebrated before the sacred fire, during the recitation of Avesta, and most of the time offerings of milk and bread accompany Haoma. It is believed that Haoma was brought to earth by divine birds. Also, Haoma is used as a medical wonder for fertility.

Fire Temple / Tower of Silence: Zoroastrians are focused mainly on Good Words, Good Thoughts, and Good Deeds, and pay very little attention to ritual worship. They pray five times a day for pure Zoroastrian waters, rivers, land, and atmosphere, and therefore, they pray anywhere they see fit. Most of the communal festivals are seasonal and celebrated in the open. From the earliest archaeological sites discovered to date, Zoroastrians had no temple or any religious structures up until the fifth century BC.

Fire Temple: One of the oldest fire temples is known as Atashkadeh-e Yazd, or Yazd Atash Behram, meaning "Victorious Fire," located in Yazd province in Iran. The Sasanian Empire constructed this fire temple; this mighty flame burned for almost 1,500 years for ceremonial purposes. The highest grade of fire can be placed inside a temple, drawing its sources from sixteen various types of fire, including from cremation pyre and lightning. The Islamic Empire completely eradicated Zoroastrianism by defeating the Persian Empire. Presently, there are about two hundred fire temples around the world.

Zoroastrians are not fire worshippers; they believe fire represents Ahura Mazda's light and wisdom. Fire burning predates Zoroastrianism; Indo-Aryans used fire burning for success in military action and other purposes long before Zoroastrians.

Tower of Silence: Zoroastrians believe the dead body is unclean, and they do not want the uncleaned corpse to pollute pure earth, fire, water, and plants.

Zoroastrians build a Tower of Silence (Avesta: Dokhma, meaning to burn from the sun) away from urban areas, at the bottom of mountains or hills. The construction of the towers is relatively uniform. It consists of a flat roof with the perimeter slightly higher than the center, and is divided into three concentric rings. The corpse of a man is placed on the outer ring. The corpse of a woman is laid on the second circle, and the corpse of a child on the innermost ring. These corpses are exposed to the sun and birds of prey. They leave the body out until the bones are bleached by the sun and wind. When the bones are bleached, they dump them in a hole and add calcium carbonate to speed up the disintegration.

Influence of Zoroastrianism on Judaism, Christianity, and Islam

There is no doubt that Zoroastrianism influenced all three Abrahamic religions. In general, the Zoroastrian impact made itself known in many ways, specifically by loaning words and concepts to the integration of religious ideas and mythic details.

In 597 BC, Chaldeans captured Jerusalem, taking thousands captive, and exiled them to Babylon, where they settled. Half a century later, the Persian Empire, Cyrus the Great, invaded Mesopotamia and liberated the Jews. In the Old Testament of the Bible, Cyrus the Great's name appears and is praised.

The following items didn't exist in the Hebrew Bible until after the Jews' liberation.

Paradise: In the Indo-Aryan language, Avestan, Paradise, is called *pairidaeza*. It is composed of two words: *pairi*, meaning around, and *daeza*, meaning wall. Pairidaeza then means "a wall enclosing a garden." The Jews borrowed paradise beginning 500 BC, when Yehud, the Jewish province, was a Persian province. In Genesis, Paradise (Greek: Paradeisos) translates as the Garden of Eden.

Heaven and hell: In Zoroastrianism, heaven is called the House of Song, filled with pleasure and happiness; hell is called the hellish abyss filled with discomfort and darkness. The Hebrew Bible had no concept of heaven and hell. After the liberation of the Jews, around 500 BC, Jews included heaven and hell in the Hebrew Bible.

Day of Judgment: Zoroastrianism believes in the final judgment (Avestan: Frashokereti, making wonderful) at the end of time. Before the liberation of the Jews, the Hebrew Bible had no concept of a day of judgment; after around 500 BC, Jews included the day of judgment in the Hebrew Bible.

Resurrection: Zoroastrians believe all souls are resurrected and reunited with the resurrected earthly bodies. The Hebrew Bible did not have any concept of resurrection until after 500 BC.

Messianic prophecy: Zoroastrians believe in the arrival of a final Messiah or Saoshyant at the end of time (Avestan: one who brings benefit). The Hebrew Bible had nothing about messianic prophecy or a Messiah; after around 500 BC, Jews included in the Hebrew Bible the concept of messianic prophecy, similar to Zoroastrianism.

The Persian Empire was the greatest empire in the world, with a highly advanced civilization, and they had a considerable political, cultural, and religious influence over the Arabian Peninsula for more than one thousand years. Muhammad's great-uncle, a leading chief of the Quraish tribe, traveled to the Persian Empire to acquire a trade agreement. Islam borrowed and modified the following items from Zoroastrianism, the religion of Persians:

Five prayers a day: Zoroastrianism has five prayers a day. Therefore, some historians argue that Islam borrowed the concept of five prayers a day from Zoroastrianism. They could have adopted this concept in two ways: during the invasion of Persia (636 CE) or during the control of the Persian Empire of the Arabian Peninsula.

Chinvat Bridge: Chinvat Bridge is a Zoroastrian concept after death, depending on the person's performance in this world. The souls who followed the falsehood will fall off the bridge into a hellish abyss. The souls who developed a good state of mind will be accompanied by a beautiful feminine, called Daena, to Paradise. Islam borrowed this concept with minor modification and called it As-Sirat, the hair-narrow bridge every human must pass on the day of judgment to enter Paradise.

After death: Zoroastrianism believes in the final judgment (Avestan: Frashokereti, making wonderful) at the end of time. Islam borrowed the final judgment from Zoroastrians and modified eternal hell or heaven. For the Zoroastrian, those who followed falsehood served in hell, and at judgment day, everybody goes to live in Paradise.

Free will: Ahura Mazda (Wise God) gave humans free will and made them morally responsible for their actions. Christianity, borrowed the concept of free will and adapted it to their holy scriptures.

Summary

Zoroastrianism is believed to be the first world's oldest monotheistic religion, and its founder is a prophet named Zoroaster (Persian: Zarathushtra). Zoroaster was one of the most exceptional and creative leaders and reformers of his time. While Zoroaster's neighboring societies, like the Indian subcontinent and the Middle East, believed in thousands of gods and goddesses and had a caste system, Zoroaster, with his creative mind, came up with a modern idea of one God.

After years of peacefully encouraging people to his new idea of one God, he was able to convert only his cousin and a few more people. Finally, he approached King Vishtaspa in Bactria (Balkh). The king and the Queen Hutaosa had heard of the religious debate between Zoroaster and the religious leaders of the kingdom. First, the religious leaders jailed him at the order of the king. Soon, the king's favorite horse got sick; King Vishtaspa knew Zoroaster knew about curing his horse, as Zoroaster's father raised horses.

After Zoroaster healed his horse, King Vishtaspa and Queen Hutaosa converted to Zoroastrianism, and they made it the official religion of the kingdom.

Regardless of how progressive a religion is, it has some roots in primitive faith. In Zoroastrianism, it is fire and the sun.

Fear is the easiest tool or method to keep people under tight control by leaders and religious authorities. In Zoroastrianism, there are two primary components of fear after death.

1. Heaven is called the House of Song and is filled with pleasure and happiness.
2. Hell is called the hellish abyss and is filled with discomfort and darkness.

CHAPTER 6

Judaism

Biblical Moses leads the Israelites through the desert Sinai during the Exodus, in the wilderness, in search of the promised land with the Ark of the Covenant, 3D render painting (Copyright iStock by Getty Images/credit: Ratpack223 and Cloudio Ventrella).

Judaism is the third oldest world religion, and the oldest of the Abrahamic faiths. Judaism traces its beginning back to about 2000 BC. Abraham (Hebrew: Avraham), whose original name was Abram (Hebrew: Avram), was born in 2001 BC (Gen. 12:4). He is the crucial figure of Judaism, and the first Hebrew patriarch to receive a revelation from God.

According to the Old Testament (Gen. 23), Abraham and his family's burial place is the Cave of Machpelah (Hebrew: Me'arat ha-Makhpela) in Hebron. Still, there is no actual proof of his or his family by archaeologists.

He is revered by three major world faiths of Middle Eastern religions: Jews, Christians, and Muslims. Abraham, son of Terah, was a descendant of Noah's son Shem and was of a Semitic race living in the city of Ur (today's Tall al Muqayyar, Iraq) in Mesopotamia among Sumerians. He was familiar with the religion and culture of Sumerians, with their thousands of deities. According to the Bible, God revealed himself to Abraham, directed him to leave his homeland in Ur, and travel where God would lead him (Gen. 11:31, 12:9). Abraham trusted this newly revealed God and made a covenant (deal) with him that all the males of his family would be circumcised. His name changed from Abram to Abraham, and his wife's name changed from Sarai to Sarah (Gen. 17:1–27).

Abraham journeyed from Mesopotamia to Egypt, and from Egypt back to Shechem, what is now southern Israel. He and his son Isaac laid the foundations of a monotheism, a new idea that could get better attention than the gods of the Sumerians. Abraham's grandson Jacob fathered twelve sons and a daughter. Jacob's twelve sons became the ancestors of the original twelve tribes of the nation of Israel.[78]

Moses, the man who led the nation of Israel out of Egypt, was from the tribe of Levi. Many centuries after Abraham, the Jews were living as slaves in Egypt. The pharaoh adopted Moses around 1576 BC (Exod. 7:7); he became the leader of Jews and a prophet. A significant part of the Jewish scripture is occupied by Moses, more so than any other single figure, starting from the beginning of Exodus to the end of Deuteronomy.

The same God who helped Abraham settle in the promised land (Canaan) helped Moses lead the Jews out of slavery and led them to the promised land. When they reached Mount Sinai, God revealed himself to Moses at the burning bush and told him his name, YHWH (Yahweh), and gave him the Ten Commandments (Exod. 20:2–17). God made the same covenant with Moses as he did with Abraham.

The Ten Commandments are the following:

1. You shall have no other gods before me.
2. You shall not carry the Lord your God's name in vain.
3. You shall not make idols.
4. Remember the Sabbath day to make it holy.
5. Honor your father and mother.
6. You shall not murder.
7. You shall not commit adultery.
8. You shall not steal.
9. You shall not bear false witness against your neighbor.
10. You shall not covet.

Jewish tradition (Torah) dates Abraham's birth to 2001 BC and Moses's birth to 1576 BC, but stories of Genesis say that Moses authored the entire Torah. There are strong beliefs among researchers that the Torah was not written until the Jews' liberation (600–500 BC).

Common Beliefs in Abrahamic Religions

There are many common facets in Abrahamic religions, and among them are Abraham exemplifies obedience, faith, and submission to God; belief in one God; God created all existence; God created angels and human souls; God created all human beings in his image and from one soul; God speaks through his prophets only; human beings must live according to the laws based on revelation; Adam and Eve were the first man and woman God created; and God instructed Noah to build an ark and caused a worldwide flood. Some of these beliefs are part of Mesopotamian hymns.

Mesopotamian story of creation: The story of creation comes from the time when humankind transitioned from hunter-gatherer to the agricultural revolution. Until this time, the gods, instead of man, did the work on the earth. The assembly of gods approved of sacrificing one young god. Geshtu-e, a pure young god who had intelligence, was slaughtered in their meeting, and a soul came into existence from the god's flesh. The goddess Mami proclaimed it as his living sign. The soul would exist so that none would forget the god.

Geshtu-e's blood and tissue were mixed with clay, and the great gods spat spittle upon the clay. Mami, with the help of the wise god Enki, created seven males and seven females.

The god Enlil was thrilled with the work of wise goddess Mami and his brother, wise god Enki. The newly created human replicated, wise Mami gave them new spades, new picks, and led them two by two down to earth to relieve the gods of their labor.

This story is the first version of creation, and later was modified by Abrahamic religions under Adam and Eve (Gen. 1:10, 12, 18, 21, 31).

Mesopotamian story of the flood: The gods grew restless as the noise of humankind became too great, and the god Enlil cried, "I am losing sleep!" Enlil organized his assembly of gods, and he expressed his frustration with the mortals. The assembly approved a flood to wipe out all of humanity.

Wise Atrahasis was looking for his lord Enki day and night, but he could not find him. Enki came to Atrahasis in a dream, and instructed him to go to the temple, put his ear to the wall, and listen to his god. Atrahasis went to the temple and placed his ear to the reed wall. Enki spoke to his servant Atrahasis through the temple wall and gave the message: "Make sure you listen to me regularly. Dismantle the house of God. Build a boat with its material. Make an upper deck and lower deck, roof it like Abzu, so that the sun cannot see inside it! Leave all personal possessions and put aboard the seed of all living things, a wealth of birds, and a hamper of fish. The bitumen and tackle must be robust to give strength to the boat. The sand needed for the flood should be seven nights' worth."

Wise Atrahasis gathered up the elders at his door. He told them that Enki and Enlil had become angry with each other. "I can no longer stay in Shuruppak. I cannot remain in Enlil's land. I must go to Abzu and stay with my beloved god. I must build a boat to take me there. That is what he told me." The elders heard Atrahasis's speech; they called the carpenters and all types of workers to his aid. Everything needed for building the boat was brought to him, and the boat was built according to the plan.

All manner of life was loaded aboard the boat. Atrahasis selected the best of all species, like cattle, wild beasts, wild animals, and all kinds of birds, and placed them on the boat. He put all his family and friends aboard the boat. He invited all his people for a feast. They were celebrating, eating, and drinking, but Atrahasis was going in and out and placing the decks of the boat. Bitumen was brought to him, and he sealed the door with it.

For seven days and seven nights, the torrent, the storm, and the flood came on. Bodies clogged the rivers like reeds. When the seven days passed, the torrent ceased, and silence fell upon the earth. The sea became calm, and the floodplain became flat like a roof. The sun came out and began to shine.

When Atrahasis heard the silence, he opened a window in the boat and saw that the sun was shining. He released a dove. The dove came back, for no perching place was visible to it. He took a raven and released it from the boat. The raven never returned.

Slowly the water receded, and the boat of Atrahasis came to rest atop Mount Nimush. Atrahasis put down the door of his boat and let out all his kith and kin and all the animals and his people within the boat. Atrahasis made a sacrifice of thanks to the god Enki, his beloved god.

This story is the first version of the great flood, and later it was modified by Abrahamic religions and replaced Atrahasis with Noah's ark (Gen. 6–8).

Leaders

Judaism traces its roots back to the most prominent and mythical individuals who first explained the tenets of Judaism to the people, and they include the patriarchs and biblical leaders.

Abraham: Abraham is the first Hebrew to receive a revelation from God. He is the one all Jews can trace their identities to, as described in the book of Genesis in the Bible. This God is all-powerful as the creator and sustainer of the universe. He is omnipotent, omniscient, omnipresent, and omnibenevolent. This God had a covenant with Abraham to leave his own country (Mesopotamia) for another land and required him to circumcise all the males in his household. Abraham completely trusted in God, and agreed to travel where God would take him.

Abraham journeyed from Ur to Haran (present-day Turkey), and from Haran, at the age of seventy-five, he settled at Shechem or Nablus, thirty-five miles north of Jerusalem, the promised land of Canaanites.

God promised Abraham in Genesis 15:4 that he would have an heir of his own body. Sarah was barren, but when Abraham was 100 years old and Sarah was 90 years old, she did conceive and gave birth to a son, Isaac. Abraham circumcised himself at the age of 99, his son Ishmael 13 years old, and his son Isaac 8 days old (Genesis 17:9–14, 24–25, and 21:4). Abraham lived for 175 years.

Abraham was a mythological character created by a brilliant biblical storyteller. According to this story, (see Gen. 18:1–15), the Hebrew god came down to Mamre to discuss and lunch under the tree's shade across from Abraham's tent. They had an excellent lunch consisting of bread, yogurt, milk, and roasted meat. After, the Hebrew god, leaving, said to Abraham, "I will return to you next year at this time, and Sarah will have a son." Sarah laughed to herself and said, "How could a worn-out ninety-year-old enjoy such pleasure, especially when my husband is one hundred years old?"

Besides, the Hebrew god made a covenant with Abraham when he was ninety-nine (Gen. 17:1–14). This covenant consisted of (1) "I make you the father of many nations." (2) "I change your name from Abram to Abraham." (3) "I will always be your god and the god of your descendants." (4) "I will give you the entire land of Canaan." (5) I tell you, your descendants and your male slaves must be circumcised. This covenant is inferior in content from a god who is supposed to be omnipotent, omniscient, omnipresent, and omnibenevolent.

Several questions arrive from an analysis of this covenant. Why did Abraham's god want to be only Abraham's god and his descendants although he was not the god of the world's population? What was the significance of requiring circumcision only for Abraham, his descendants, and his slaves while most of the world's population do not circumcise? Can a woman at ninety conceive from a man who is at one hundred? What omnibenevolent god would test Abraham by demanding that he sacrifice his child, Isaac, and then suddenly place a ram from nowhere under his knife? None of the above be explained by science and reason. There is no known archaeological or independent research confirming the existence of Abraham.

Isaac: Isaac (Hebrew: Yitzchak) was the promised son to Abraham (Gen. 17:15–27). He tested Abraham, asked him to sacrifice Isaac on an altar as a burnt offering. Abraham followed God's wish, and as he was about to cut Isaac's throat, God provided a ram from nowhere to sacrifice instead. Jews revere Isaac as he was removed from the altar, firstborn to Jewish parents, and lived all his life in the promised land of Israel, and he was the second patriarch of Jewish people.

Isaac married Rebecca, the daughter of Abraham's nephew Bethuel, at the age of 40. Rebecca was barren, and they were praying to God; she conceived after 20 years of marriage with twins, Jacob, then after his death, Esau. A famine ravaged the promised land. Isaac decided to go to Egypt, but God intervened and told him not to because Isaac acquired holiness when he was placed on the altar during binding. Therefore, he could no longer leave the sanctity of the promised land for a land of less holiness. He lived 180 years of his life in Canaan.

Jacob: Jacob is the third patriarch of Judaism and the great patriarchs of the Old Testament. Jacob was bright; sometimes this trait worked for him, and sometimes it backfired on him. Jacob's life is mostly tragic. His brother, Esau, wanted to kill him, so he fled to his uncle Laban in Haran, his mother's hometown. He fell in love with Rachel, the youngest daughter of his uncle, but his uncle tricked him into marrying her oldest sister, Leah. Jacob was at the age of eighty-four. Later, he married Rachel. Jacob had twelve sons, from which the entire Jewish populations descend, and a daughter, by his two wives, Leah and Rachel, and by their handmaidens, Zilpah and Bilhah.

After many years, while Jacob was heading back to Canaan, he had an all-night wrestling match with God, demanding his blessings. In the end, God touched Jacob's hip, and from that night on, he walked with a limp. God blessed him and renamed him Israel, the one who "wrestles with God."

Later, after the occupation of Canaan by Jews, the promised land was subdivided into twelve tribes, and Jacob's twelve sons became the ancestors of the twelve Jewish tribes. Jacob lived for 147 years.

Moses (Hebrew: Moshe): Moses's date and place of birth are not known. It is accepted that in 1400 BC he was considered one of the most important religious leaders and a prophet who founded Judaism.

Goshen is the name of the land where Hebrews were living in biblical times. By the time of the Exodus from Egypt, they were leaving from the eastern delta of the Nile.

The Hebrew population was growing, and the pharaoh was afraid they would revolt and overthrow him. He ordered the death of all newborn Hebrew males. Moses was born during this time, so his mother hid him for three months, and then she made a basket and put the baby Moses in it and laid it at the riverbank. The pharaoh's daughter, Bithiah (or Bitya), came to the river to wash, saw the baby floating in the basket, and told the maids to fetch it. Bithiah adopted the child and named him Moses (Exod. 2:1–10). The future Jewish leader was raised as an Egyptian prince in the court. He trained in Egyptian military, civil, and governmental matters.

The Torah records three incidents in Moses's life before God appointed him a prophet.[79]

1. He saw an Egyptian overseer beating a Jewish slave, and he killed the overseer.
2. The next day, he tried to make peace between two Jews who were fighting, but the aggressor said to Moses, "Do you mean to kill as you killed the Egyptian?"
3. The pharaoh ordered Moses killed, and he fled to Midian. Moses married Tzipporah, a daughter of a Midianite priest, and became the shepherd for his father-in-law's flock.

While Moses was in the wilderness of Midian with the flock of his father-in-law, he saw a burning bush (Exod. 3:2–3). God was inside the burning bush and told Moses that he wished him to go to Egypt and confront the pharaoh, to lead the Hebrew slaves out of Egypt, and take his people to another land that was flowing with milk and honey.

Moses balked, but when God insisted, Moses said to the pharaoh, "Let my people go!" The pharaoh, as expected, refused Moses's demands, but a few miracles occurred that finally caused the pharaoh to change his mind.

First: As a punishment to the pharaoh, God sent ten plagues.

1. The Nile River turned to blood.
2. Frogs infested the land.
3. Lice came upon the families.
4. Wild animals were everywhere.
5. Cattle died of disease.
6. The people were infected with boils.
7. Strong hails came and destroyed almost everything.
8. Swarms of locusts descended and ate everything left after the hail.
9. Three days of darkness ensued
10. Every firstborn in Egypt died. (The Israelites were protected from this by smearing lamb's blood on their doorframes so the Lord would pass over their homes and they would be safe.)

None of the plagues affected the pharaoh's decision except the last one. As a result, the pharaoh agreed and let the Israelites go.

Second: As the Israelites were heading toward the promised land, the pharaoh changed his mind and sent his army to bring them back. The Egyptian army chased the fleeing Israelites to the bank of the Red Sea (Sea of Reeds). God sent a mighty east wind that moved the high seawater to the sides and made a path of dry land in the middle. The Israelites passed through on the dry path, but when the Egyptians followed them, the water walls collapsed, and the entire Egyptian army drowned.

Third: Moses, with God's people, wandered for forty years in the wilderness between Egypt and Canaan. The people of God were thirsty and demanded that Moses supply them with water. God commanded Moses to assemble all the Hebrews and, in front of them, order a rock to yield its water. Moses angrily told his people, "You rebels, shall we get water for you out of this rock?" He banged the rock twice with his rod, and water came out of the rock, as God had promised. Moses did not follow all God's commands, though, and this was the reason God prohibited him from entering the promised land (Num. 20:10–11).

Fourth: Two months after the Israelites departed from Egypt, they were directed to be prepared for meeting the God of Israel at Mount Sinai (or Mount Horeb; it is believed that Mount Sinai and Mount Horeb are the same mountain). The people were terrified by the Lord's appearance with massive thunder, rain, and lightning. Moses reassured them that God would never hurt them, but with this appearance, the fear of God must stay in them so they would not go astray. Moses ascended Mount Sinai for forty days and nights; God revealed himself to Moses and gave him the Ten Commandments and the rest of the Torah. That is why Moses is considered the most important prophet of Judaism and a significant figure in Christianity and Islam.

Scientists have discovered the earliest Hebrew writing dating around 1000 BC, during the reign of King David. Scholars agree that the Old Testament, or the sacred scripture of the Jewish faith, was written at different times and completed around 160 BC.

God of Moses, who was omnipotent, omniscient, omnipresent, and omnibenevolent, could change the Egyptian pharaoh's mind to let Moses and his people be free to go to the promised land of Canaan. This god did not do that, but he sent to the empire ten plagues and sent a mighty east wind to build a high-water wall on both sides of the Red Sea and a path of dry land for Israelite to pass through. The Egyptian army following the Israelites to capture them at this dry path, the water-wall collapsed, and the entire Egyptian army drowned. None of the above are mentioned in the history of the Egyptian empire.

Judaism

The omnipotent and powerful God freed Moses with his people from slavery and let them enter the wilderness between Egypt and the promised land. In two months after Moses and his people departed from Egypt, this God revealed himself to Moses at Mt. Sinai, gave him the Ten Commandments and the entire Torah. After this startling event, Moses and his people wandered for forty years in the wilderness. His God prohibited Moses and his people from entering the promised land (Num. 20:10–11). Moses died at the age of 120, and it is not known where he died and how.

All the above said cannot be explained by science and reason. God gave him Torah while Hebrew writing was discovered around 1000 BC, and Torah was completed around 160 BC. Moses was a mythological character created by a brilliant biblical storyteller. Other than the Bible stories, there is not a shred of archaeological or independent research confirming Moses's existence.

King David, 1000 BC (1 Sam. 17): David was a musician, poet, hymnist, and a warrior-king. He was the second king of Israel. He is a key character in the Christian Bible, and is known to be "a man after God's own heart." For all his failings he loved God, and is thought to have written most of the psalms included in the Bible.

During the reign of King Saul (the first king of Israel), the Israelites were being subjected to Philistine domination and bullying. David offered himself as a warrior against Philistinian warrior Goliath the giant, who stood nine and a half feet tall and carried a lance, sword, and spear. King Saul offered David his armor and sword, but he shrugged them off and responded, "They are too heavy. Instead, I need a stick, a sling, and five smooth stones."

David called on God to help him and killed Goliath instantly with a stone to the forehead, then lopped off the giant's head. He presented it to King Saul, and the Philistinian troops fled in panic.

David started his career at the court of King Saul. As a warrior against the Philistinians, his popularity aroused Saul's jealousy, however, and Saul wanted to kill him. David fled to the south, to the coastal plain of Palestine, where he started to plan his future career.

Soon after the Philistines killed King Saul on Mount Gilboa, David was invited to become the successor of King Saul. He defeated all enemies and built a vast empire. He accepted Jerusalem as the capital of the people and the religious center of the Jews. He united all the Jewish tribes by marrying a wife from each tribe. Jewish scriptures foretell that a descendant of David, the Messiah, would restore the kingdom of Israel.

Was King David a prophet? Some believe he was and some do not. His name is mentioned many times in the Old and New Testament, as well as in the Koran as a prophet. In real life, he was famous and influential; he did not need to be a prophet for people to listen to him.

King Solomon (1 Kings 1–11): Jewish tradition considers Solomon (Hebrew: Shlomo) King David's son. He was crowned around 967 BC while King David was alive, even though he was younger than his brothers. Solomon was the author of three biblical books: Proverbs, with three thousand proverbs; Song of Songs, with over a thousand songs; and Ecclesiastes.

Solomon commenced his reign after his father, King David, who had given him the vastest and securest kingdom in Jewish history. Solomon was the first to build the Temple of Jerusalem (the first Beit ha-Mikdash, "Great Temple"). The Temple of Jerusalem was destroyed during the Babylonian siege of Jerusalem in 586 BC. Solomon was revered in Judaism and Christianity for his wisdom, and in Islam as a prophet.

Solomon was the most polygamous in Jewish history. He had seven hundred spouses and three hundred concubines. Solomon built idolatrous temples for his non-Jewish wives to have places to worship.

Hillel the Elder: Hillel was born in Babylon in 110 BC and died in Jerusalem in 10 BC. He was a very famous Jewish scholar and sage in Jewish history. He was a poor man and tried hard to become a rabbi. He was found on the skylight at the House of Learning, where Rabbi Sh'mayah and Avtalyon were talking about living. Hillel became the successor of Rabbi Sh'mayah and Avtalyon.

Hillel is credited with a lot of insightful proverbs: "One who is shy will never learn," and "If I am not for myself, who is for me? And when I am for myself, what am I? And if not now, when?"[80]

The Torah had legislated that personal debt was to be canceled every seventh year (Deut. 15:1–2). This legislation could hurt the poor and would only be for personal liability, not court debt. Hillel created a procedure called *prosbul*. It was beneficial for business and the economy. The lender had to note before the court that he would collect his debt. Hillel was involved in the development of the Talmud (Schools of Hillel and Shammai). He was often contrasted with Shammai; on many occasions, the disputes between Hillel and Shammai were on a ritual basis, with the decision in favor of Hillel made on moral ground.

Shammai: Shammai was a contemporary to Hillel the Elder, and both were working in the development of the Talmud. The Schools of Hillel and Shammai are repeatedly mentioned in Talmud stories.

Golda Meir: Golda Meir (original name Goldie Mabovitch) was born on May 3, 1898, in Kyiv, Ukraine, and died December 8, 1978, in Jerusalem. Golda Meir's family moved to Milwaukee, Wisconsin, USA, in 1906. She graduated from the University of Wisconsin and later became a leader of the Labor Zionist Party in Wisconsin. On March 17, 1969, she became Israel's fourth prime minister, until 1974. After her death four years later, she was buried at Mount Herzl in Jerusalem.

There are many other significant figures, like Maimonides or Rambam, David Ben-Gurion, Menachem Mendel Schneerson, Rabbi Akiva, Moses Mendelssohn, and more. As earlier discussed, Jews were on the move from Abraham's era to May 14, 1948 (the date the UN declared Jerusalem a sovereign state),[81] searching for a better place for comfort and survival.

Scriptures

The Torah is the central and most important document of the Jewish Bible and a divine revelation to Israel (the Jewish people). Jews believe that the Torah tells how God wants them to live. The meaning of the Torah is the law of God as revealed to Moses and written down in the first five books of the Hebrew scriptures. These are the books traditionally attributed to Moses, the recipient of the original revelation from God on Mount Sinai. Jewish people, Christians, and Muslims agree that Moses was the author of the five books. The three Abrahamic religions agree on their order: Genesis, Exodus, Leviticus, Numbers, and Deuteronomy.

The written Torah, the first five books of the Bible, is handwritten on parchment scrolls that reside inside the ark of the law—preserved in all Jewish synagogues. There is strong belief among researchers that the Torah was not written until 600 BC; scientists have discovered the earliest Hebrew writing dating from 1000 BC, during the reign of King David.

The Tanakh is an acronym of three Jewish books in the Bible or Old Testament. Torah means laws, Nevi'im means prophet, and Ketuvim means writing. Tanakh tells the whole story of God's relationship with the Jewish people.

The Talmud is an essential body of Jewish civil and ceremonial laws and legends comprised of the Mishnah, the first part of the Talmud (a written collection of orally transmitted traditions). The second part is Tosefta and the Gemara (additional commentary on the Mishnah). The Talmud takes all the rules in the Torah and explains how to apply them in different circumstances. The Talmud is in two versions: the earlier Palestinian or Jerusalem Talmud and the Babylonian Talmud.

The Mishnah is the first section of Talmud and is an authoritative collection of biblical exegesis embodying the oral tradition of Jewish law and the first part of the Talmud. It was published in the second century CE and deals with agriculture, sacred times, marriage, damages, holy things, and purity laws.

The Halakah means in Hebrew "the path that one walks." Halakah is also spelled Halakha or Halacha. The most sacred of the Halakah is the 613 commandments that instruct followers, among other things, "not to do work on Sabbath," "not to eat the meat of unclean beasts," and "to honor the old and the wise."[82]

There are several other scriptures in Judaism outside the scope of this book, like the Midrash, the Mishnah Torah, the Haggadah, and more.

Denominations

Judaism is the third oldest religion in the world and has been active for over 4,000 years. The total population of Jews around the world is slightly more than 14.5 million as of 2017. Judaism, much like other dominant religions, is divided into several denominations around the world; every denomination is devised and led by a rabbi or leader.

However, the divisions of Judaism active today are not the same as seen in the Bible. Therefore, the ancient and modern times have to be understood separately. It isn't in the scope of this book to give a complete description of every denomination.

The historical function of a leader in Judaism: Abraham was a patriarch, Moses was a prophet, David was a king, Golda Meir was a prime minister, and all were the Israelites' leaders.

Four thousand years ago, Abraham and God talked to each other and made a covenant. The agreements were

- Israelites are God's people;
- Abraham had to leave his country;
- Abraham and his household males must be circumcised;
- Abraham's wife, Sarah, got pregnant at the age of ninety for their promised son Isaac; and
- Abraham must sacrifice his son Isaac, and as soon as he raised the knife in obedience to God, his son was replaced with a ram.

Three thousand years ago, King David, a strong fighter and a leader, possibly a prophet, invaded Jerusalem and created for the first time in history a mighty Israelite empire.

Fifty years ago, Golda Meir was not a patriarch nor a prophet; she was a competent leader. She recreated an Israel state in the promised land. She didn't need to convince the Labor Zionist Party of her prophecy; she needed the support of the United States and European powers. How long Israel will be in control of its own destiny, history will tell.

Division of Judaism in the Ancient Era

Pharisees Judaism: Pharisees were the first denomination in the Jewish religion; they are the ancestors of all contemporary Jews. Pharisees believed in Oral Law, and they thought that Oral Law empowered them with change based on need.

Sadducees Judaism: The Sadducees were opponents of the Pharisees. Sadducees were part of a wealthier denomination; the majority of them were rabbis who were serving at the temples. They rejected the Oral Law of the Pharisees; they followed the Bible and dismissed the notion of the afterlife because it was not mentioned in the Torah. The Sadducees went out of existence after the destruction of the temple in 70 CE.

Essenes Judaism: The Essenes were the third denomination; they were like an ascetic and disciplined group of ancient hippies. Essenes didn't have any desire to be part of the temple; they believed the Sadducees corrupted the temple. They did not have private houses, and they lived together and pooled their income.

Zealots Judaism: Zealots are the fourth denomination; they arose at the beginning of the common era. The Roman Empire ruled the Israelites and installed high priests at the temple. Jews revolted against the Roman Empire in 60 CE. During this time, Zealots' ranks grew. Zealots believed that the Israelite's God would fight alongside the Jews. By the end of 68 CE, around one hundred thousand Jews were killed or sold into slavery, and destruction of the temple would end the Zealots' power.

Beliefs

Why do the Abrahamic religions believe in one God? All three faiths have the same heritage, culture, shared roots, and they were part of the great civilization of Mesopotamia. Mesopotamian civilization lasted for thousands of years. They believed and worshipped multiple deities, which were assembled into a pantheon of goddesses and gods. These deities became old and were not able to protect their believers from natural disasters, poverty, drought, and starvation. Abraham was part of it. His father, Terah, was an idol maker. Abraham destroyed all idols, which angered his father. It was time for a change from polytheism (multiple deities) to monotheism (one God).

Many creedal formulations came at different times in Jewish history. None lasted long or was accepted other than the confessions of the oneness of God with a single sentence in the Shema (Deut. 6:4): "Hear, O Israel: The Lord our God, the Lord is One," or, "The Lord our God is one Lord."

Mesopotamia was invaded by the Persians, Cyrus the Great, in 539 BC. Cyrus the Great let the Jews go back from the Babylonians' captivity to Jerusalem. Cyrus allowed them to reconstruct the temple in Jerusalem. Therefore, Cyrus the Great's name appears in multiple places in the Old Testament.

Zoroastrianism, the religion of the largest empire in the world, had an enormous impact and influence on Judaism, Christian, and Islam. The Abrahamic religions borrowed many beliefs from Zoroastrianism, including one God, Paradise, heaven and hell, day of judgment, resurrection, messianic prophecy, five prayers a day, Chinvat Bridge, soul, afterlife, and free will.

To the Jews, their single God is represented by four Hebrew letters: YHWH. This God is all-powerful as the creator and sustainer of the universe; he is omnipotent, omniscient, omnipresent, and omnibenevolent. The Jews' relationship with God is a covenant relationship. Their religion's backbone is the Torah, comprised of five books of the Bible, which was the whole of the laws and narrative from the beginning of the world to the death of Moses as given to the Jews in Mount Sinai. They must follow God's rules daily. Besides the Torah, they include the Nevi'im, or the books of prophets.

Creation: The story of creation is found in the first section of the Torah. During six days, God created first the heavens and the earth, then he created the light, sky, the seas, and plants, fish, birds, and animals. Humans were made on the sixth day, and God rested on the seventh day (Gen. 1:1). "So God created human beings in his own image. In the image of God he created them; male and female he created them" (Gen. 1:27).

Funeral: The Jewish funeral consists of a burial, also known as an interment. Family members prepare the body based on tradition, and burial ideally occurs within a day of death. Cremation is forbidden. Sitting shivah[83] refers to seven days of mourning for the family members. Friends and relatives visit the home of the deceased to extend their sympathy.

Resurrection: Jews were kind of silent about the resurrection. In the Old Testament, there are a few passages about a future bodily resurrection. On what happens after resurrection, the Jews are silent. "Many of those whose bodies lie dead and buried will rise, some to everlasting life and some to shame and everlasting disgrace" (Dan. 12:2, Isa. 26:19). Also, they name the resurrection of the dead in the Amidah, or standing prayer.

Afterlife: The afterlife (Hebrew: Olam Ha-Ba) is seldom discussed among Orthodox, Conservative, or Reform Jews. The Torah, or Old Testament, has no explicit reference to the afterlife. In the Islam and Christian religions, the afterlife plays a critical role, especially in Islam; those who die in jihad (English: holy war) will ascend to the highest place in heaven. Therefore, Muslim terrorists take the suicide missions seriously.

Soul: Before the liberation of Jews from Babylonian captivity in 539 BC by Cyrus the Great, there was nothing about the concept of an immaterial and immortal soul separate from the body in Judaism. After 539 BC, with the influence of Zoroastrianism, the idea of the soul developed in Judaism. Added to the Bible are these words: "For then the dust will return to the earth, and the spirit will return to God who gives it" (Eccles. 12:7). At a different place in the Bible it says, "Then the Lord said, 'My Spirit will not put up with humans for such a long time, for they are only mortal flesh. In the future, their normal lifespan will be no more than 120 years'" (Gen. 6:3).

Festivals and Holy Days

Holy days and festivals are the most crucial part of the Jewish faith. The Judaism calendar is a lunisolar structure. It is regulated by the position of both the moon and the sun. There are 12 alternating lunar months of 29 and 30 days each. Leap years were introduced to conform to the solar year of 365.24 days. The first month of the year is that in which the Exodus began. The Jewish calendar reflects Jewish history and its teaching.

Rosh Hashanah is the Jewish New Year. It occurs around the middle of September or October.

Yom Kippur occurs shortly after Rosh Hashanah, around September or October. It is the Day of Atonement. On this day, Jews seek purification, and it is the time to repent for sins of the past year. This day runs from sunrise to sunset, and Jews must abstain from food, drink, and sex.

Sukkot occurs in September or October. Sukkot is an autumn festival that celebrates a joyful harvest. This festival is also known as the Feast of Tabernacles, or the Feast of Booths. This holy day lasts a week, and people build huts called Sukkot. They eat in there and sometimes sleep.

Hanukkah (Chanukah) occurs late November to mid-December. Hanukkah is officially a minor festival but recently has been celebrated more than significant festivals. Hanukkah is known as the Celebration of Lights. It marks the victory of the Maccabees over the Syrians in 200 BC.

Purim occurs late February to early March. Purim commemorates the deliverance of the Persian Jews from annihilation at the hands of ancient Persians. The day before Purim is spent fasting, and the Jewish children dress as the characters in the biblical stories and it is a day of joy.

Pesach (Passover) occurs from late March to early April. Pesach is a weeklong observance, and it honors the delivery of the Jewish people from slavery. During this celebration, Jewish people eat unleavened bread, known as matzoh. It is commemorating the quickly made unleavened bread the Jews had to survive on during their escape from Egypt.

Shavuot occurs in May to June. Shavuot translates into Greek as "Pentecost." It lasts for two days. It is the spring harvest festival and commemorates God's revelation of the Torah to Moses.

Rituals and Customs

Given the fact that Jewish history goes back to 2000 BC, over four thousand years, a significant number of customs, rituals, and traditions have been amassed. In Judaism, religious observances and rituals are noted in Jewish law, or Halakah, which means in Hebrew "the path that one walks." Many Jews have a *mezuzah* on every doorpost in the home to remind everyone to keep God's law.

Sabbath: Sabbath (Hebrew: Shabbat) means "cease." It is a day of religious observance and abstinence from work, kept by Jews from Friday sunset to Saturday nightfall. According to the biblical book of Exodus, the Sabbath is a day of rest on the seventh day of creation, as God rested on the seventh day, and it is commanded by God to be kept as a holy day of rest.

God miraculously provided the fleeing Israelites manna, "bread from heaven" every Friday. Therefore, the Jews were not forced to gather food on the Sabbath for forty years while wandering in the wilderness.

Brit Milah, or the circumcision ceremony: This covenant was between Abraham and his God to circumcise all the males in his household. It is the ritual for the removal of the foreskin of the penis, which is performed following Genesis 17:10. It is supposed to take place on the eighth day after birth, while also giving a name to the boy. The circumcision is accomplished by a specialist called a *mohel*, who is often a rabbi, in the baby's home.

Brit Hayyim / Brit Bat: In ancient times, there was no ceremony for newborn girls. Brit Hayyim is a naming ceremony for baby girls. It also takes place on the eighth day after birth.

Bar Mitzvah: A formality for boys at the age of thirteen, it marks a Jewish boy's entry into the community as an adult. The word means *son of the commandments*.

Kiddushin: The Jewish marriage ceremony. It takes place under a canopy called a huppah and includes the ritual breaking of a glass underfoot. The breaking of the glass is an act to recall the destruction of the Jewish temple in 70 CE.

Temple / Synagogues / Shul of Jerusalem

Jewish congregations were known as holy assemblies (Hebrew: kehillot) as from the Greek translation, synagogue means assembly. In Eastern Europe, a synagogue is called a shul, which in Hebrew means school. With the emergence of the reform movement in the first decade of the nineteenth century, the first temple was established in Hamburg, Germany.

After Moses died around 1406 BC, Joshua took over, and finally, the Israelites invaded Canaan. Canaan was divided into city-states. In those days, the Ark of the Covenant had no permanent place and moved among several sanctuaries.

King David captured Jerusalem in 1004 BC. He selected Mount Moriah for the temple, where it is believed Abraham had built the altar on which to sacrifice his son Isaac. This temple was the first permanent house for the Ark of the Covenant. The temple was built and completed by King David's son Solomon.

Nebuchadnezzar II of Babylonia destroyed the temple in 587–586 BC. Cyrus II, founder of the Achaemenian Dynasty of Persia, allowed the Jews to return to Jerusalem and rebuild the temple after the conquest of Babylon in 538 BC. Construction of the temple was completed in 515 BC. During the fourth and third centuries BC, the temple generally was respected.

The Jewish rebellion against the Roman Empire began in 66 CE. Soon after this rebellion, the Roman Empire focused on the temple and destroyed it on 70 CE. The only part of the temple that remained was the portion of the Western Wall, which is called the Wailing Wall, which is a focus of Jewish pilgrimage and aspiration.

After the rise of Islam, Jerusalem was invaded by Muslims under Umar ibn al-Khattab, in 637 CE. The Dome of the Rock and Al-Aqsa (Arabic: Qubbat al-Sakhrah) shrine in Jerusalem was built by the Umayyad caliph Abdul al-Malik ibn Marwan in 691 CE; it was returned to Jewish control in 1967.

Division of Judaism in the Modern Era

Orthodox/Hasidic Judaism: During early modern Europe in Germany, a group under the leadership of Samson Raphael Hirsch called themselves Orthodox Jews (anti-enlightenment). Orthodox follows the authority of the halakhic (legal) system and traditional Jewish laws and customs, not only the liturgy, but also diet and dress. Those laws include separation of the sexes during worship. A group within Orthodox Jews was founded by Rabbi Israel ben Eliezer, called Hasidic, also known as Baal Shem Tov. Hasidic cut themselves off from the modern world to pray and study the Torah. All Hasidim are Orthodox, but not all Orthodox are Hasidic. The majority of Orthodox Jews are presently living in Israel.

Reform Judaism: The Reform movement started in Germany under the leadership of Israel Jacobson in the early nineteenth century. He was seeking to integrate Jews into mainstream society. They changed their diet, and moved their Sabbath from Saturday to Sunday. Jacobson's liturgy was in German rather than Hebrew. Most of them believe in human authorship rather than the divine origin of the Torah. They were able to be accepted into German society without having to convert to Christian belief.

Conservative Judaism: The Conservative movement started under the leadership of Rabbi Zachariah Frankel (1801–1875) in Breslau, Germany. Conservatism tries to conserve essential elements of traditional Judaism but allows for the modernization of religious practices in a less radical sense than the Reform movement. In 1980, the Conservatives decided to admit women as rabbis.

Reconstructionism Judaism: The Reconstructionist movement started under the leadership of Rabbi Mordecai M. Kaplan (1881–1983) in the United States of America. This denomination adapted classical Judaism to the thinking of science, reason, and art. He started the first Bat Mitzvah, conferring on young women a religious rite of passage previously reserved only for males. Reconstructionists see Judaism as an evolving civilization rather than a religion. Reconstructionists reject the notion of a personal God, and miracles like a transcendent God who made a covenant with his chosen people, parting of the Red Sea, and do not accept the Bible as the inspired word of God.

Humanistic Judaism: Rabbi Sherwin T. Wine founded Humanistic Judaism in 1963 in Birmingham Temple, Detroit, Michigan. The temple grew to 140 families by 1965 and over 400 by the time of his death in 2007. Humanistic Judaism practices the nontheistic form of Judaism. Judaism is the culture and historical experience of the Jewish people; history has taught Jews to rely on human power to discover the truth. Humanistic Judaism does not use theistic language in its liturgy. It focuses on the ethnic culture of the Jewish people, which includes participation in traditional Jewish holidays and all life events.

Morality

Morality has nothing to do with belief in God and religion. Morality is hardwired in the human brain and makes the distinction between right and wrong; morality is an evolved faculty with a genetic basis.[84]

It is clear from observation of the Golden Rule of various religions that moral codes are meant to minimize the suffering of others. Among all religions, Jainism (see below), which does not believe in God, is ranking the highest in morality.

The following is a review of the universal moral codes of various religions, usually called the Golden Rule (do unto others as you would have them do unto you), in chronological order.

- Hinduism: The oldest organized religion is Hinduism, which goes back to the Indus Valley civilization (2300–2000 BC). In the Mahabharata (Sanskrit: Great Epics of the Bharata Dynasty) book 13, section CXIII, verse 8 says, "One should never do that to another which regards as harmful to one's self, and this called the rule of dharma. Other behavior is due to selfish desires."
- Zoroastrianism: Zoroastrianism is the world's oldest monotheistic religion (1500–1200 BC) and its founder,

Zoroaster, says, "Do not do unto others whatever is injurious to yourself."
- Judaism: Judaism is the oldest Abrahamic religion. In the Old Testament, Leviticus 19:18 says, "Do not explore revenge or bear resentment against a fellow Israelite, but make sure love your neighbor as yourself."
- Jainism: The founder of Jainism is not known. Mahavira (~599 BC) is the most regarded Jain, and gave Jainism its present-day form. They avoid the suffering of any living thing, and they respect the environment and do not believe in God. The sacred scripture Agamas says, "Just as sorrow or pain is not desirable to you, so it is to all who breathe, exist, live or have any quintessence of life. To you and all, it is undesirable and repugnant."
- Buddhism: Buddha was born in Nepal in 563 BC. Buddha said, "One who, while himself seeking happiness, oppresses with violence other beings who also desire happiness, will not attain happiness hereafter" (Dhammapada 10). He also said, "Hurt not others in ways that you would find yourself hurtful" (Udanavarga 5:18).
- Taoism: The founder of Taoism, Lao Tzu (or Laozi) was born in 601 BC in China. "The sage has no interest of his own, but take the interests of the people as his own. He is kind to the kind; he is also kind to unkind: for virtue is kind. He is faithful to the faithful; he is also faithful to the unfaithful: for virtue is faithful" (Tao Te Ching, chapter 49.)
- Confucianism: Confucius was born in Shandong in 551 BC. He did not believe in God, and he was a teacher. He said, "Do not impose on others what you would not choose for yourself."[85]
- Christianity: There are a few mentions of the Golden Rule in the Christian Bible. "Do to others whatever you would like them to do you" (Matt. 7:12). "Do to others as you would like them to do to you" (Luke 6:31). "Do not seek revenge or bear a grudge against a fellow Israelite, but love your neighbor as yourself" (Lev. 19:18).

- Islam: The Arabs considered the survival of tribes paramount, and the ancient rite of blood vengeance ensured them. Muhammad was from the same culture as the survival of the tribes. In the Koran, there are several scriptures of moral guidance and religious law. "But in Hadith, a Bedouin came to the prophet, grabbed the stirrup of his camel, and said: 'O the messenger of God! Teach me something to go to heaven with it.' Prophet said: 'As you would have people do to you, do to them; and what you dislike to be done to you, don't do to them. Now let the stirrup go! [This maxim is enough for you; go and act by it!]'" (Kitab al-Kafi, vol. 2, p. 146).
- Sikhism: Sikhism religion and philosophy sprang from Hinduism and Islam. The Guru Granth Sahib says, "If thou desirest thy Beloved, then hurt thou, not anyone's heart" (Guru Arjan Dev Ji 259, Guru Granth Sahib).
- Bahaism: The Tablets of Baha-Ullah says, "Blessed is he who preferreth his brother before himself" (Baha-Ullah).

Summary

Judaism is the third oldest world religion and the oldest of the Abrahamic faiths. Judaism traces its beginning back to about 2000 BC. According to the Old Testament of the Bible (Gen. 23), Father Abraham (Hebrew: Avraham), whose original name was Abram (Hebrew: Avram) was blessed by God as the progenitor of the twelve tribes of Israel.

Moses is the second most significant leader and prophet of Judaism. Two months after the Israelites departed from Egypt, they were directed to be prepared for meeting the God of Israel at Mount Sinai (or Mount Horeb). Moses ascended Mount Sinai and stayed there forty days and nights. God revealed himself to Moses and gave him the Ten Commandments and the rest of the Torah. Moses and God's people wandered for forty years in the wilderness between Egypt and Canaan and suffered much. God did not allow Moses to enter the promised land because Moses disobeyed him.

Judaism did not progress and expand because their myths and stories did not give them incentive other than Israelites are God's people (Judg. 20:2, 2 Sam. 14:13, Isa. 44:1, Gen. 35:12) and the promised land described in the Old Testament (Gen. 12:1, 15:18–21, 26:3, 28:13, Exod. 23:31). It is clear that essential leaders from Abraham to Moses were mythical individuals, and the Torah or Old Testament are unverifiable stories. Besides, the stories were disseminated through the centuries by word of mouth.

In general, religion evolved based on fear-emotion. Let's see if the element of the fear exists in Judaism. Fear is the only tool or method to keep people under control by the leaders and religious authorities. Jews believe death is natural, and after death the person will be disconnected from God, earth, and family. During the time of Moses, Jews were looking for the life on earth. Korah rebelled against God (Num. 16:11), and he was in front of Moses. The earth was opened and consumed Korah and his associates and families, including their possessions. It is the real element of fear from God.

In the later dates, especially after the liberation of Jews by Cyrus the Great and influence of Zoroastrianism, new items were inserted into the Jewish scripture like soul, afterlife and resurrection, and more. The two significant components of fear and reward to control people in Judaism are as follows:

- *The Garden of Eden or heaven*: The truly righteous like sages, prophets, and kings after death, their souls ascend to the Garden of Eden.
- *Sheol or Underground Abyss, also called Gehinnom*: The average people after death their souls descend in Sheol is described in the Bible as a region dark and deep or a place of punishment or purification.

CHAPTER 7

Jainism

In the Osian Jain temple, Rajasthan, India, a stone carving of a Jain in repose (Copyright iStock by Getty Images/ credit: Nityajacob and Greveshkovmaxim).

Jainism is one of the religions specific to the Indian subcontinent. All the Indian subcontinent religions believe in karma and rebirth, and they respect nature and revere their rivers, mountains, trees, animals, all living creatures, and earth. All religions in the Indian subcontinent have common beliefs, like historical, literary heritage, and a tradition of asceticism. Hindus and Jains worship images and share a single temple for worship.

Jainism emerged as a nontheistic religion; they believe there are no gods that will help human beings, or they have an ethics-based faith, and also as a reaction against the teaching of Orthodox Brahmanism, based on a caste structure.

Jainism was founded in far more ancient times. As a continuation in the development of Jainism, in eastern India in the sixth century BC, a young prince named Jina Vardhamana Mahavira, as the twenty-fourth and last Tirthankara (Sanskrit: ford maker), a significant perfected historical figure, acted as a teacher in the search for perfection. Mahavira abandoned the palace, his wife, children, and all his possessions at the age of thirty. He started on a life of extreme asceticism in search of moksha, or mukti (Sanskrit: liberation), from the Hindu's continuous cycle of karma and rebirth where the immortal soul lives forever in a state of bliss.

Mahavira spent twelve years meditating; he wore the same clothes for one year, and after that, he discarded his clothes and walked naked. He strictly followed the principle of ahimsa (Sanskrit: nonviolence). He endured the physical hardship from society and the nature of his ascetic life peacefully. He accepted free food in the hollow of his palm. After twelve years of this severe lifestyle, he finally attained Kevala Jnana (Sanskrit: Absolute Knowledge) or omniscience and bliss.

Hindus came to accept Jain asceticism and ahimsa (Sanskrit: not to injure). It started as a small movement, and remains so today, mostly confined to the Indian subcontinent. Possibly the strict requirements of Jainism kept their religion as a minority religion, with four million followers in the world.

Leaders

The founder of Jainism is not known. It is believed Jainism was founded in far more ancient times. There are twenty-four founding Tirthankaras that are essential leaders to the Jains; some critical leaders are from the Hindu tradition, like Krishna and Rama, are listed below.[86]

Krishna is revered by the Jains as a cousin of the twenty-second Tirthankara, Arishtanemi, also called Neminatha.

Rama was a hero to the Jains and was treated by them as a pious and nonviolent Jain.

Rishabhanatha also known as Rishabhadeva, Rsabhadeva, or Rsabha (Sanskrit: Lord Bull), he was the first Tirthankara (Sanskrit: ford maker), and the first Jina (Sanskrit: a liberated great teacher), and he was the first to preach the Jain faith. He is a mythical leader. Jainism believes Rishabhanatha lived many millions of years ago. Legend credits him with teaching the men a couple of primary professions, specifically swordsmanship, writing, agriculture, knowledge, trade, and commerce, as well as teaching the women many crafts, such as weaving, pottery, and carpentry. Rishabhanatha attained moksha on Mount Kailas in the Himalayas, the birthplace of the Hindu Shiva. Rishabhanatha is said to have taught seventy-two sciences, which include visual arts, arithmetic, singing and dancing, and the art of lovemaking.

Bahubali (or Gommateshvara) is a very revered figure among Jains and especially among the Digambara sect of Jainism, who believe Bahubali was the first person to reach moksha. He was the son of the first Tirthankara, Rishabhanatha. After Bahubali won a duel with his half brother Baharata for control of the kingdom, he refused to kill Baharata. He gave up his clothing, possessions, renounced the world, and became a monk. Legend says he stood immobile, meditating, for an entire year; vines grew up around his body from feet to head. His meditation allowed him to attain his liberation and Kevala Jnana (Absolute Knowledge).

Several works of sculpture depict Bahubali, including the fifty-seven-foot-high statue of him in the town of Shravanabelagola, India, a center for the Digambara sect. It is one of the world's largest freestanding statues. Every twelve years, his figure is a pilgrimage site for Jains. The entire sculpture is ceremonially bathed in milk, curd, and ghee before enormous crowds of people.

Parshvanatha,[87] also known as Parshva or Paras, was the twenty-third Tirthankara (Sanskrit: ford maker) of Jainism. Parshvanatha was the first Tirthankara with historical evidence; he was living from 872 to 772 BC. Prince Parshvanatha, at the age of thirty years, renounced the world to become a monk. He meditated for eighty-four days, and he achieved moksha at the age of one hundred years. Parshvanatha established the fourfold restraints: (1) not to take life, (2-) not to steal, (3) not to lie, (4) do not own property. (And with Mahavira's addition of the vow of celibacy, these became the five great vows of Jain's ascetics. See below.)

Mahavira is the last Tirthankara. The Jain texts state he was born in 599 BC into the royal family of Kshatriyas in Bihar, India. He is regarded as the one who gave Jainism its present-day form; he is sometimes (wrongly) called the founder of Jainism.

His childhood name was Vardhamana, which means prosperous, or one who grows. He had all the luxuries as a prince and grew up to be a courageous young boy, and once brought a very fierce serpent under his control. This bravery earned him the name Mahavira (Sanskrit: Great Warrior). He received the title of Jina (Sanskrit: the victorious). He added the fifth vow of the Jain's ascetics, celibacy, so the five great vows are ahisma, not to take life; satya, not to lie; asteya (or acharya), not to steal; aparigraha, do not own property; and brahmacharya, celibacy or chastity.

There are many other vital leaders in Jainism, but it is not within the scope of this book to list them all.

Scriptures

Jainism's sacred text authenticity is disputed between the sects of Svetambaras and Digambaras. The holy books contain Mahavira's teaching, called Agamas, which are the canonical literature, the scripture of Svetambara Jain. Mahavira's disciples compiled his teachings into texts and memorized them to pass on to future generations.

The texts had to be learned because Jain monks and nuns were not allowed to possess religious books as part of their fourth vow to not own property. As time passed, some of the texts were forgotten, misremembered, additional commentaries were added to the texts, and particularly the twelve years of continuous famine around 350 BC killed off many of the Jain monks who had memorized the texts.

The Svetambara sect believes that the majority of the Agamas texts survived, except the fourteen Purvas were forgotten. Therefore, they decided to document the Agamas as remembered by them at various times.

The Digambara sect believes that during the twelve years of famine, all the Agamas were lost. Therefore, Digambara, in the absence of authentic scriptures, uses two primary texts and commentaries on the main texts. They are four Anuyogs consisting of more than twenty texts as the basis for their religious practices. Great scholars wrote these scriptures.

Denomination

Jains are divided into two major denominations, or sects; they are the Svetambara, or white-clad sect, and the Digambara, or sky-clad sect. Each of these denominations is further broken into sub-denominations or subsects. Both factions agree on the basics of Jainism and disagree on the detailed life of Mahavira, the spiritual status of women, and which texts should be accepted as scripture. One of the significant differences where they separate from each other is the concept of Aparigraha, which means do not own property.

Svetambara or white-clad sect: Svetambaras accept the authority of Agama, the Holy Scripture or Texts. Svetambara monks and nuns wear simple white clothing. Svetambaras believe that men and women are equally capable of achieving liberation and moksha, but in real life, monks play a more important role than nuns. Svetambaras do not go to such extremes as the Digambara in the name of purity and austerity, and they believe in favor of women, unlike the Digambara sects. The nineteenth Tirthankara Lord Mallinatha, named Malli Bai by the Svetambaras, is believed to be a woman.

One of the notable and significant differences between the two sects is the concept of Aparigraha, which means do not own property. Svetambara monks can own five possessions:

1. A danda (Sanskrit: a staff)
2. A whisk to remove insects from their path
3. A bowl for begging for food
4. An item of simple white clothing with a square mask on their mouths to prevent the accidental inhalation of the insects
5. A book with writing material

Svetambara monks wander during the dry season in the countryside near temples and near sites of pilgrimage, begging for food. Their images and statues of Svetambaras are created to be visible to viewers, and they have prominent staring eyes.

Subsects of Svetambaras are Muti Pujaks, who worship in the temples; the Sthanakvasi and Terapanthi, both of which emphasize ahimsa, or nonviolence, to its extreme.

Digambara or sky-clad sect: The most austere monks and nuns of Digambara take the virtue of aparigraha, which means do not own property, in a very extreme sense. They follow Mahavira's way of life, and they believe a perfect divine should possess nothing, not even a piece of clothing. Therefore, the monks and nuns of the Digambara denomination prefer nudity. They are more rigid in their observances toward all aspects of divineness, and consequently, they maintain celibacy. Digambara sects believe that women are not capable of achieving moksha, or liberation, unless they are reborn as a man. Because women cannot live an ascetic life because they cannot live naked, it is inappropriate because sexual excitements and women are naturally harmful.

The only possession a Digambara monk and nun can possess is a whisk made of peacock feathers, which they use to sweep the insects from their path of the walk, and a gourd to store drinking water. Digambara Jains reject the authority of the agamas and use four anuyogs consisting of more than twenty texts as the basis for their religious practices. The Digambara images and statues of Tirthankaras are plain, and have downcast eyes and are always carved as naked figures.

Subsects of Digambara are Bisampantha, who worships the idols of Tirthankaras with saffron, flowers, fruits, and sweets; and Terapantha, who worship the idols of Tirthankaras without saffron, flowers, fruits, and desserts, and the name Terapantha appears in both the Svetambara and Digambara sects. Terapantha focuses more on sacred literature and spiritual values.

Belief

Jains do not believe in God; they have an ethics-based religion. Jains do believe in divine beings who are worthy of devotion.

Creation: Jains believe the universe always existed and will live forever, and there was no being or force to create the world.

Divine beings: Jains venerate Jinas, who have achieved infinite perfection and have become liberated from the cycle of birth and death.

Tirthankaras: Tirthankaras are a group of twenty-four religious divine who attained moksha and became Siddha (Sanskrit: perfected one or pure soul). Jains believe they are divine in the hope that they will be encouraged by them along the path toward their final liberation.

Samsara: Samsara is a worldly cycle of birth, death, rebirths, and reincarnations in various realms of existence, like plants, animals, humans, and nonliving things. Samsara is nonending, as long as the soul is in bondage to its karma. Jains believe moksha is the only way to be liberated from Samsara.

Karma: Karmas (Sanskrit: actions, deeds, or works, including thoughts and words) are invisible fine particles of matter all over the universe, just like air particles. These particles are attracted to the Jiva (soul) by the action of Jiva. The Jiva is constrained to a cycle of rebirth. This cycle will end with an individual's effort and intention to achieve moksha or liberation.

Moksha: Moksha, also spelled moksa (Sanskrit: free, liberate, let go, release), is the spiritual liberation from the cycle of Samsara.

According to the principle of Jainism, this liberation will occur when an individual follows the three jewels in Jaina, or Ratnatraya,[88] and they are the right belief, right knowledge, and proper conduct.

Siddha: Siddha (Sanskrit: perfected one or pure soul) is a pure soul who has removed himself or herself from the cycle of Samsara by right belief, right knowledge, and proper conduct. This soul resides in a state of infinite bliss, in the Siddhashila, at the top of the universe. In Jainism, a Siddha is revered and divine.

Asceticism: Asceticism in Jainism means austerities or body modification, and in Hinduism means inner cleansing to self-discipline. Central to Jainism to achieve moksha is asceticism, or self-discipline. Many of Jainism's ascetic applications can be traced back to Jina. Vardhamana Mahavira's five great vows are the following:

1. Ahimsa, not to take life
2. Satya, not to lie
3. Asteya or Acharya, not to steal
4. Aparigraha, do not own property
5. Brahmacharya, celibacy or chastity

Jains' strict asceticism influenced other religions like Christianity, Judaism, and Islam.

Death and the afterlife: Death and the afterlife in Jainism could end up in a variety of ways; it is dependent on a person's karma. After each bodily death with bad karma, the soul (Jiva) is reborn into a different body to live another life, and this process will continue until it achieves liberation. They do not rule out the possibility that a soul may become reborn in Naraka's place of torment.

Jains call the spiritual ritual of dying Sallekhana. In older age, terminal illness, or by gradually reducing the amount of food intake until death—as did Chandragupta Maurya, the founder of the Mauryan Empire—they are practicing Sallekhana. They read scripture, then prepare for a peaceful passing, saying something like, "May all beings forgive me for any harm I may have done to them, intentionally or unintentionally." Jains call Sallekhana a peaceful purification from karma.

The Five Great Vows of Jainism

The five great vows of Jainism are undertaken by ascetics and accepted as fundamental to Mahavira's teaching by both Svetambaras and Digambaras. It shows the path and advises the Jain followers to achieve their purity, and they follow the road voluntarily.

Jainism does not give any commandment or diktat. It only shows the path and advises its followers to follow that path to their maximum capability. Thus, followers of Jainism do not have to take any vows. The five great vows are ahimsa (not to take life), satya (not to lie), asteya (not to steal), aparigraha (do not own property), and brahmacharya (celibacy or chastity).

Festivals and Holidays

Jains celebrate their religious holy days by worshipping, reciting sacred texts, giving and getting alms, and fasting. The festivals play an essential role in the Jains' community and show their devotion. Participation is not obligatory.

Diwali is usually celebrated annually in October or November throughout India as a famous Hindu festival. Jains call it the Festival of Lights, and they light candles. It was a special day when Mahavira attained moksha, or nirvana.

Mahavir Jayanti is the celebration of Mahavira's birthday, which falls in late March or early April and is observed by both Jain sects, the Svetambaras and Digambaras. Jains gather in temples while images of Mahavira are paraded in the streets and crowds recite specific chants and prayers.

Mahamastakabhisheka festival is held every twelve years to honor Lord Bahubali, or Gommateshvara, a very revered figure among Jains and especially among the Digambara sect of Jainism. He is the son of the first Tirthankara, Rishabhanatha. This festival takes place in Sravanabelgola in Karnataka, south India. His fifty-eight-foot-high statue is a pilgrimage site for Jains. Every twelve years, the entire sculpture is ceremonially bathed in milk, curd, and ghee before enormous crowds of people.

Paryushana is a year-end festival that is observed at the end of August or September. Paryushana is one of the oldest Jain festivals and is celebrated by both sects, eight days by the Svetambaras, and ten days by the Digambaras. It is the time for fasting, taking vows, and to keep their minds firmly fixed on religion. They go to temples and worship the Tirthankaras, which are perfected beings. Jains are encouraged to spend time as ascetics and at least twenty-four hours of meditation with monks.

Kartik Purnima is celebrated on the full moon day of Kartik (around November). The sacred hill surrounding Palitana is closed during the rainy months. Jains believe that this is the area where the first Tirthankara, Rishabhanatha, meditated before his first sermon. The community starts eating vegetables, and the area reopens for business. Many people start on a pilgrimage to the hills and Temple Palitana.

Rituals and Customs

Jains believe there is no God or gods that will help human beings on the road to liberation. They believe that human life is several births, deaths, and rebirths, to the point that the soul has shed all karma and achieved liberation. In order to achieve this liberation, they must follow the three ethical jewels, which are right belief, right knowledge, and proper conduct. Their main rituals and customs include the following:

Prayer: Daily, Jains bow their heads and say their universal prayer, Navkar Mantra (also Namaskar Mantra, and Namokar Mantra) as follows:

- I bow to the Arithantas—the ever-perfect spiritual victors.
- I bow to Siddhas—the liberated souls.
- I bow to Acharyas—the leader of the Jain order.
- I bow to Upadhyayas—the learned preceptors.
- I bow to all saints and sages everywhere in the world.

These fivefold praises destroy all bad karmas and all of the holies. It is the foremost of the holy.

By saying this prayer, Jains do not ask for any favor or material benefit from their divines, Tirthankaras, monks, or nuns. Jains salute them to receive the inspiration from the five benevolents for the right path of real happiness and freedom from the misery of life.

Weddings: In Jainism, marriage is recognized as a worldly affair. The Jainism scriptures emphasize the dependence of two living organisms on each other. There are some ritual differences between the two main sects of Jainism. The Digambaras have a grand total of twenty prescribed rituals, while the Svetambaras have a total of sixteen rituals. A Jain wedding begins seven days before the ceremony and seven days after the ceremony and includes some rituals and other ceremonies.

The wedding ceremony is performed under a mandap, which is a four-post canopy. These four posts symbolize the four parents, who worked very hard to raise their children. The couple sits beneath the canopy with their parents. The priest will perform a few rituals and congratulate the couple on their marriage and give the blessing.

Worship: Most Jains have a shrine at home. They bathe and dress simply in two pieces of cloth. They bow and recite the Navkar Mantra before the image of Siddha and pass three times around the image. They wash the picture with water and milk and a mixture of sandalwood and saffron.

The pious Jains go daily or occasionally to the temple for worshipping the image of Tirthankara. They take a bath and wear simple cloth. If they see idols of Tirthankara, they have to say Namo Jinanam and place folded hands over the bent forehead, which means Solute Jina.

Jain Temples

The Indian subcontinent is really a melting pot of several beautiful cultures, traditions, and religions. Jainism is one of them. Jains built magnificent temples with incredible architectural beauty for their Tirthankaras across the Indian subcontinent.

Dilwara Jain Temple: This temple is situated at the only hill station of Rajasthan—Mount Abu—and was constructed during the twelfth century. The interior of this structure has remarkable carvings over marble, and its hall has 360 miniature statues of Jain Tirthankaras.

Khajuraho Temples (Shantinath): The Shantinath temple is located in the eastern part of the state of Madhya Pradesh, India. This structure may have been constructed in 1870 CE, but the pedestal of the 12-foot Shantinath image bears an inscription dated 1027–1028 CE. The temple appears from the hill at 3 levels and is supported by around 1,444 beautifully carved marble pillars, and one single marble rock having 108 heads of snakes and a tangle of tails.

Palitana temples: Palitana temples are located on Shatrunjaya Hill by the city of Palitana in the Bhavnagar district of Gujrat, India. These temples, dedicated to Rishabhanatha, who is also known as Adinath, meaning the first lord, was the first of the Jain Tirthankaras. These temples have over three thousand years of history of Jainism. The earlier temples were built under the patronage of King Kumara Pala of the Solanki Dynasty in the eleventh century CE. These complexes were destroyed and ravaged by the invading Islamic armies of Allaudin Khilji in the thirteenth century CE. In 1656 CE, Mogul emperor Shah Jahan's son Murad Baksh granted Palitana immunity, and Emperor Akbar helped complete the restoration of Palitana.

Gomateshwara Temple: Gomateshwara Temple is located in the town of Shravanabelagola, in the state of Karnataka, in India. Gomateshwara, or Bahubali, was the second son of the first Tirthankara, Rishabhanatha. His fifty-eight-foot-tall monolithic statue was built around 938 CE, and this statue can be viewed at about a distance of nineteen miles.

Summary

Jainism is one of the religions specific to the Indian subcontinent. The founder of Jainism is not known. Jainism emerged as a nontheistic religion; they believe there are no gods that will help human beings. They have an ethics-based faith, also as a reaction against the teaching of Orthodox Brahmanism, which is based on a caste structure. They avoid causing suffering of any living thing, and

Jainism

they respect the environment. Mahavira is the most regarded Jain, who gave Jainism its present-day form in 599 BC.

Jain's strict asceticism influenced other religions, like Christianity, Judaism, and Islam.

In Jainism, like every other major religion, there are two primary components.

1. *Karmas* (Sanskrit: actions, deeds, or works, including thoughts and words) are invisible fine particles of matter, all over the universe, just like air particles. These particles are attracted to the Jiva (soul) by the action of Jiva. The Jiva is constrained to a cycle of rebirth.
2. *Moksha*, also spelled Moksa (Sanskrit: free, liberate, let go, release), is the spiritual liberation from the cycle of Samsara. According to the principle of Jainism, this liberation will occur when an individual follows the three jewels in Jaina, or Ratnatraya, and they are the right belief, right knowledge, and proper conduct.

BUDDHIST

CHAPTER 8

Buddhism

Old Buddha statue meditation religious symbol at roots (Copyright iStock by Getty Images/credit: GummyBone and Coolvectormaher).

Buddhism, like Jainism, is another one of the religions specific to the Indian subcontinent. The Indian subcontinent religions have the same culture. They believe in karma, rebirth, reincarnation, meditation, and they respect nature. They also have common beliefs like history, a tradition of asceticism, and literary heritage.

Buddhism emerged as a nontheistic religion, and there is no belief in a personal god. Buddhists believe that whatever power the gods might possess, it could not solve the human dilemma. The path for awakening or enlightenment is through the practice of meditation, morality, and wisdom. Some scholars don't recognize Buddhism as an organized religion but rather a way of life or a spiritual tradition.[89]

Buddhism is a pathway of spiritual development leading to the attainment of a deep insight into the true nature of life. According to Pew Research Center 2012,[90] there are 520 million practicing Buddhists in the world, making Buddhism the fifth largest of the world's religions. Buddhism is the majority religion in Thailand, Cambodia, Japan, Laos, Mongolia, Myanmar (Burma), Sri Lanka, Thailand, Bhutan, Singapore, and Vietnam.

Buddhism, like Zoroastrianism, spread throughout the world without imposing its beliefs by force or the killing of innocents like other religions, such as Christianity and Islam did. Buddhism has always placed great emphasis on nonviolence concerning all life. The first precept in the Pali Canon is to abstain from taking life. Therefore, the communities of monks and nuns, known as bhikkhus and bhikkhunis, traveled along the road to spread Buddha's teaching without imposing force.

Like other major religions, Christianity and Islam, Buddhists never wrote down Buddha's teachings. Buddha's disciples memorized his words, and their followers carried on oral tradition because the entire original teaching was easy and short to remember. The first comprehensive written scripture of the Buddha, including additions and commentaries by monks and abbots, was compiled about five hundred years after his death.

Buddha

Siddhartha Gautama, or in Poli, Gotama Buddha, Siddhartha (Sanskrit: He Who Achieves His Goal), Gautama (Sanskrit: descendants of Gotama; Gotama is the name of several prominent figures in Rig Veda), the Buddha (Sanskrit: Awakened One or Enlightened One), was born to a life of privilege—into a royal family of the Kshatriya, or warrior, in Lumbini—in present-day Nepal in 563 BC. Seven days after his birth, his mother died. His father was King Suddhodana, the leader of the Shakya clan. His royal family was Hindu, in the warrior caste, and Gotama Buddha was raised and learned as a Hindu all the deities, cultures, and traditions, and died as a Hindu.

There are stories or myths that Gotama Buddha possessed superhuman powers and abilities, like other Middle Eastern prophets. It said that, after his birth, he stood up, took seven steps north and uttered, "I am supreme of the world, Eldest am I in the world, Premier am I in the world, this is the last birth, there is now nothing further coming to be."[91] Furthermore, everywhere baby Gotama Buddha placed his foot, a lotus flower bloomed.

He was a master of archery, and he won his wife, Yashodhara, the daughter of King Suppabuddha. He was married at the age of sixteen to Yashodhara, and together they had one child, a son, Rahula.

On several occasions, Gotama Buddha had to leave the palace to see the outside world. Poverty and leprosy were widespread in India; he could see them even around the castle. Some days, Gotama Buddha was out and viewed them from a close distance. This had a different effect. He saw an older person, a dead body, some young content ascetic, and a leper lying on the road in pain, and lots of poor people. Very likely, Gotama Buddha was impacted by what he saw, and especially the poor and the people with leprosy and diseases.

When Gotama Buddha was growing up, there was a Jainism movement in response to the teaching of Orthodox Brahmanism, which was based on the caste system. This movement understood that gods and goddesses are not able to help the poor and cure leprosy and other diseases to stop humanity from suffering. At the age of twenty-nine, Gotama Buddha came to the same conclusion.

According to accepted tradition, the twenty-nine-year-old Gotama Buddha, on the night of his birthday, left the palace for the search of knowledge, just like a doctor, to find a solution for human suffering from diseases and poverty. For six years he was wandering around. First, he went to two Hindu teachers for improving his understanding of Hinduism; he did not find a solution. Later, Gotama Buddha joined a group of ascetics, representatives of Jainism. After a couple of years of practicing Jainism, he realized that this religion's best interest is to achieve moksha. He decided that of Jainism's five great vows, ahisma and aparigraha are detrimental to human life.

Gotama Buddha moved on for the search of enlightenment and decided to meditate alone beneath the sacred bodhi tree. He decided, according to tradition, and he vowed to meditate without moving or standing from his seat until he discovered the truth. During this period (seven weeks, or forty-nine days) of meditation, Gotama Buddha went through four trancelike stages and finally achieved enlightenment. After this period, he became known as the Buddha, meaning Awakened One or Enlightened One.

On the night of his thirty-fifth birthday, Buddha is described in many texts as Shakyamuni or Bhagavat. Buddha was part of the Shakya clan and muni (Sanskrit: ascetic or monk), which means ascetic or monk of the Shakya clan. Bhagavat means lord or blessed one.

Buddha, after his enlightenment, returned to Benares and reunited with the five ascetics with whom he had parted company. His first sermon was delivered in Deer Park, Sarnath, and set out the doctrine of the Four Noble Truths. He defined the Middle Way, which is accessible in everyone's experience, in this sermon, or Buddha dharma.

In India during that time, there were two major topics of daily discussion: Vedic dharma or Sanatana dharma or Hinduism, and Jain Dharma. Vedic dharma emphasized the enjoyment of life, and Jain dharma emphasized self-denial, total self-control, self-torture, and self-mortification.

Buddha continued preaching for the next forty-five years. During the Buddha's travels, he returned to his birthplace in Lumbini. He met his father, wife, and son, and he apologized to his father and wife. He kissed his father's foot and said, "You belong to a noble line of Kings, but I belong to the lineage of Buddhas." When he met his wife and had an informative discussion, his wife became a nun and his son a monk.

His first cousin, Ananda, joined the crowd of Buddha as his attendant. The Buddha made a massive break in Hindu tradition, permitting women to join the assembly or community (Sanskrit: sangha), and the first Buddhist nun was the Buddha's aunt who had raised him. Buddha died at the age of eighty, the night of his birthday in Kushinagar, in around 483 BC. The cause of his death was likely food poisoning.

Disciples and Leaders

Buddha is the first leader and founder of Buddhism. Buddhism, unlike other major religions, such as Islam and Christianity, is a non-violent religion. During his forty-five years of preaching, he had only a few disciples and followers. The Buddha and his followers were under the strong influence of Brahmanism, as well as the community of monks and nuns (sangha). Buddha did not assign a successor after his death, and the Buddhists had no leader until the first Buddhist council around 400 BC. For at least two hundred years, the Buddhist monastic community was overshadowed by Hinduism; only a few communities of monks and nuns traveled along the roads and were devoted to his teaching. The essential leaders in Buddhism are as follows:

Ananda, Buddha's first cousin on his father's side, joined the crowd of Buddha as his attendant in the second year of Buddha's ministry. He attained enlightenment many years later, close to the start of the First Buddhist Council at around 473 BC. Ananda convened the First Buddhist Council and could recite all the teachings or dharma from memory, and he lived 120 years.

Mahakasyapa was called the father of the sangha. His original name was Pipphali. After Buddha met Pipphali, he recognized him as a disciple, and Pipphali recognized Buddha as a teacher. The Buddha gave Pipphali the name of a great sage, or Mahakasyapa.

After Ananda convened the First Buddhist Council, Mahakasyapa led the council and was elected as the Buddhist leader.

Ashoka the Great was the last emperor of the Maurya Dynasty. He ruled from 265 to 238 BC. Ashoka invaded the Kalinga country with a bloody conquest, and after that, Ashoka renounced military conquest and converted to Buddhism. Ashoka started to practice dharma actively, and he sometimes went out and preached dharma (cosmic law and order) to the rural people and relieved their suffering. He was building hospitals for humans and animals and supplying medicine, roads, and planting trees along the streets, and erected memorial pillars, public utilities, stupas, and shrines. During Ashoka's reign, Buddhism spread all over India, and subsequently beyond the frontiers of India, making Buddhism a world religion.

Nagarjuna lived in the second century CE. He was an Indian Buddhist philosopher. He is famous for his writing about the doctrine of emptiness (sunyata). He wrote a few sacred texts, including Mulamadhyamakakarika (the Fundamental Wisdom of the Middle Way). Nagarjuna was a strong proponent of Mahayana Buddhism.

Bodhidharma, also called Putidamo (Chinese) and Daruma (Japanese), was a sixth century CE Buddhist monk. Bodhidharma's life is legendary rather than historical. A legend states that he was not happy after repeatedly falling asleep during meditation, and therefore, he cut off his eyelids. Legend also states that Bodhidharma is credited with establishing Zen Buddhism (Mahayana Buddhism).

Kanishka the Great (Chinese: Chia-ni-se-chia), from the Kushan Dynasty, came to rule an empire in Bactria (present province of Balkh in Afghanistan) between 78 and 144 CE. His reign lasted for 23 years. Kanishka's mighty empire covered approximately 2,500,000 square kilometro (970,000 square miles), from Uzbekistan in the west to the Ganga River valley in the east and from Tajikistan in the north to central India to the south. His original religion is not known, but was very likely Zoroastrianism. The summer capital of Kanishka's empire was in the ancient city of Kapisa, located in the province of Parwan, near present-day Bagram, Afghanistan, and the winter capital was in Peshawar in Pakistan. Kanishka converted to Buddhism and, as a patron of Buddhism, convened the Fourth High Buddhist Council in Kashmir and marked it as the beginning of Mahayana Buddhism. After Ashoka, he spread Buddhism all over eastern Asia, including his vast empire and central Asia.

Avalokiteshvara (Sanskrit: Looking on the Lord) is the most popular Buddhist. Avalokiteshvara is revered and worshipped by all Buddhist denominations: Mahayana (Greater Vehicle), Theravada (Way of the Elders), Vajrayana (Diamond Vehicle), and the Tantric or Esoteric sect of Buddhism. Avalokiteshvara is a bodhisattva (Buddha of unlimited compassion and mercy); he is the most popular of all leaders in Buddhist legend. Avalokiteshvara postponed himself from Buddhahood until he had helped all human beings on earth to achieve liberation. In China, he is mostly worshipped in the form of a female.

Manjushri (Sanskrit: gentle, sweet, glory) is one of the bodhisattvas (Buddha-to-be), personifying wisdom and insight. Buddhist scripture was composed in his honor around 250 CE, and Buddhist art sometime after 400 CE. He is mostly wearing princely ornaments and with the right hand holding a fiery sword (to cut through ignorance) and in his left hand, a lotus.

Kshitigarbha (Sanskrit: Womb of the Earth) was a kindhearted bodhisattva who promised not to enter nirvana until he saved all sentient beings from falling into hell. Kshitigarbha is usually depicted as a monk with a shaved head, wearing a robe and carrying a clerical staff or khakkara to the gates of hell, and a wish-fulfilling jewel to light up the darkness. He became very popular in China as Dicang, in Japan as Jizo, and in Tibet as Sa-e Nyingpo.

Maitreya (Sanskrit: friendliness) is the future Buddha who presents a bodhisattva dwelling in the Tushita heaven. Maitreya will descend to the earth during a time when the world has forgotten the teaching of Buddhism, and he will preach new dharma or law as a successor to the present Buddha, Gotama Buddha. Maitreya is known in China as Milefo, in Japanese as Miroku, in Mongolia as Maidari, and in Tibet as Byama-pa (kind or loving).

Samantabhadra is the most worshipped bodhisattva in the Mahayana denomination and represents kindness. He protects all beings who teach the dharma. He is mostly depicted in a triad with

Shakyamuni or Buddha and Manjushri bodhisattva. He is shown riding on an elephant with three heads or with one head and six tusks.

The Dalai Lama is the head monk of Tibetan Buddhism. He was governing Tibet, and his residence was there until 1959, with the invasion of Tibet by China. He fled Tibet and lives in India. According to Tibetan Buddhism, the Dalai Lama is a reincarnation of the past lama, who decided to be reborn to complete meaningful work. This process of continually being reborn is known as *tulku* and started recently, in 1391 CE. There are a couple of ways that High Lamas may determine who is the next reincarnation. Those ways are a dream of High Lamas to locate the home of the boy; the smoke of the last Dalai Lama's cremation in which direction the boy is found; and Oracle Lake, where High Lamas go to the holy lake called Lhamo Lhatso, in central Tibet, look at the wave direction, where the village of the boy is located. Using the above process, they determined the home and village of Tenzin Gyatso, the current Dalai Lama.

Scriptures

The Buddha (Siddhartha Gautama) was a prince, and he lived a life of luxury and pleasure in his father's palace. He observed poverty and disease around the castle.

The reaction was against the caste system of Brahmanism and the gods and goddesses of Hinduism who didn't help those who suffered from diseases and poverty. The Buddha examined both the exciting social lives in India, the extreme self-gratification of Hinduism and the self-mortification of Jainism. He selected the Middle Way for the best of society.

Buddha's teachings are short and concise; every follower can memorize them, which is called dharma. The later followers, during the six centuries after Buddha's death, like in every other religion and political party, tried to take advantage and created and expanded the scriptures.

The first division known as the Hinayana or small vehicle was created as a result of differing interpretations of Buddha's existence. Among this division, the most crucial sect is the Theravada (Wisdom of the Elders); this sect is responsible for Pali Canon Scripture, and this scripture is practiced all over Southeast Asia.

The second division known as the Mahasanghikas (Proponent of the Great Assembly) was created because they thought the Buddha, in his previous life, was a bodhisattva or a Buddha-to-be, before his birth. They believed the Buddha was a cosmic spiritual being, not mortal. All but one volume of their humongous Buddhist Hybrid Scripture is lost.

The third division known as the Sarvastivadins (Proponents of the View that All Exists) was created because they highlighted the significance of bodhisattva's spiritual evolution as a model of growth in morality. Their original sacred writings were lost, and Chinese translation exists.

The fourth division known as Mahayana (Great Vehicle) was created. They believe the Buddha was more than a mere mortal, whose empathy filled the universe, whose charm was available to all who asked. Mahayana scriptures are sutras, arguing to be the Buddha's own words.

At present, there are a sizeable number of scriptures in Buddhism, and some of the major ones are as follows.

Tripitaka (Pali Canon): The Tripitaka (Three Baskets of Wisdom) is considered to be a record of the Buddha's words. The Tripitaka was written down in the first century CE, under the leadership of Emperor Kanishka, and it contains the following sections:

1. *The Vinya Pitaka*: This section includes the rules directing monks and nuns, how to behave with each other and within society.
2. *The Sutta Pitaka*: This section contains the Buddha's teachings. It includes the Dhammapada, or the collection of 423 of the Buddha's sayings in verse form. It is the best known Buddhist scripture and the most widely read. It also includes the Metta Sutta; in this scripture, the Buddha explains how a person can live a life of loving-kindness.
3. *The Abhidhamma Pitaka*: This section contains teaching about the nature of life and the reasons for being. It is the most complex division.

Mahayana Sutras: Mahayana Buddhists respect Pali Canon, but in general, follow the teaching of Mahayana Sutras. The sutras are the collections of Buddha's teachings and sayings. The sutras were written between the first century BC and the fifth century CE; some of them were written later. Mahayana Buddhists have hundreds of influential texts, including the significant sutras: The Avatamsaka, the Heroic Gate (Shurangama) Sutra; the Jewel Heap (Ratnakuta) Sutra; the Lankavatara Sutra; the Lotus (Saddharma Pundarika) Sutra; the Mahaparinirvana Sutra; the Perfection of Wisdom (Prajnaparamita) Sutra; the Pure Land Sutra; the Vimalakirti Sutra; and the Tibetan Book of the Dead. The majority of Mahayana Buddhist tradition accepts these as original teachings of Buddha, practiced all over East Asia.

Religious Festivals and Holidays

Megha Puja: This festival, also known as Fourfold Assembly Day or Sangha Day, is celebrated on the full moon of the third lunar month, Indian calendar. On this day, Buddha was speaking to a gathering of 1,250 enlightened monks (arahats).

Parinirvana Day: Parinirvana Day, also known as Nirvana Day, is celebrated by some Buddhists on February 8, the Buddha's final entry into nirvana. Some Buddhists celebrate on February 15, which commemorates the death of the Buddha when he reached nirvana at the age of eighty.

New Year Theravada: Theravada Buddhists celebrate the New Year on April 28, with symbolic elements of sand and water. In the Theravada Buddhist tradition, each grain of sand is representative of wrongdoing. That means a bad mark on one's karma. When the sand is washed away by the river or by other means, the evil deed is washed away.

Buddha Day: Buddha Day, also known as Vesak or Visak Day, is celebrated on the full moon of the old lunar month of the Vesakha, which usually falls in May or early June. On Buddha Day, Buddhists commemorate the birth of the Buddha-to-be, his enlightenment, and his final passing into nirvana. For all Buddhists, the day marks Buddha's birth.

Asala: Asala, also known as Dharma Day, is celebrated on the full moon day of the eighth lunar month of the ancient Indian calendar. Dharma Day is associated with the Buddha preaching his first sermon, just after achieving enlightenment.

Uposatha: Uposatha is a Buddhist day of observance and started from Buddha's time, and still today, the Theravada Buddhist tradition of Southeast Asia observes this day. The monks reaffirm the teaching of the Buddha, and they recite the 227-rule monastic code, the patimokkha. Lay Buddhists, wearing white robes, go to the monastery to listen to Buddhist teachings from the monks and practice meditation. Usually, they observe on the day of the full moon and the new moon; later Buddhists added on the quarter days in the lunar cycle, making four holy days each month.

There are a few more festivals and events, like *Vassa*. This annual "rain retreat" is the yearly three-month retreat observed by the Theravada Buddhists, usually beginning in July.

All Souls' festivals are a specific celebration in November; in China and Japan, it is a day of respect for one's parents, elders, and ancestors.

Denominations

Buddha, during the forty-five years of his preaching, had only a few followers. His teaching was straightforward and directed to the individual he was facing. Soon after his death, there was competition to take his place, and sectarian Buddhism was formed. Like-minded monastics clumped, taught, and divided. This process carried on for hundreds of years, and now there are three major denominations and many sub-denominations.

1. Theravada
2. Mahayana
3. Vajrayana

The Theravada (Pali: Wisdom of the Elders). The elders here mean the senior Buddhist monks. This denomination is primarily responsible for the Pali Canon. The Theravada denomination is surprisingly unified, and they are mostly in countries in Southeast Asia. Sometimes, it is called Southern Buddhism. Theravada Buddhists believe that the Buddha was a historical human being who encouraged others to follow their path to enlightenment.

Mahayana (Sanskrit: Great Vehicle) is widely fragmented all over the east and central Asia, but all under the same belief. The origins, date, and tradition of Mahayana Buddhism are unknown. These sub-denominations are Pureland (Sanskrit: Buddhasetra), Zen (Sanskrit: Dhynam), and Yogacara (Sanskrit: Yoga Practice). Tathagatagarbha doctrine means the womb of the absolute and the essence of Buddhahood. Tiantai means the platform of the sky. Madhyamaka is the doctrine of the middle way. Trikaya (Sanskrit: three bodies) is part of Mahayana sect.

1. The Dharmakaya (the Body of Essence), the supreme state of absolute knowledge, the unity of all beings and things
2. The Sambhogakaya (the Body of Enjoyment), the heavenly mode
3. The Nirmanakaya (the Body of Transformation), the natural way, or emanation body.

Vajrayana (Sanskrit: Thunderbolt Vehicle, or Diamond Vehicle). This Buddhism sect arose from Mahayana. Vajrayana is also called Tibetan Buddhism, and bases its teachings on the Tantric or Esoteric sect of Buddhism. It means an esoteric stretch without a break of Buddhist thought and practice for a quicker path of enlightenment. Vajrayana has the following subsects: Dzogchen (great perfection), Gelug, Kadampa, Kagyu, Nyingma, Sakya, Jonang, and Rime.

Beliefs

Dharma: Dharma is the doctrine and teaching of Buddha, and he declared in his first sermon, "I teach suffering, its beginning, ending, and path of release."[92] They are the Four Noble Truths as follows:

1. *The First Noble Truth of Suffering*: Life is full of suffering (Dukkha). Pain comes in many shapes and forms, like

sickness, old age, and death. It seems eternal, but Buddha's teaching tells how to end suffering.
2. *The Second Noble Truth of the Cause of Suffering*: The cause of suffering (Samudaya) can be identified as hunger, pain from an injury, and thirst. The Buddha thought the root of suffering is desire (tanha) and described it as the Three Roots of Evil, or the Three Poisons, or the Three Fires, and they are greed, ignorance, and hatred.
3. *The Third Noble Truth of the Cessation of Suffering*: The cessation of suffering (Nirodha), the Buddha thought, is to overcome desire, to liberate oneself from attachments, and attaining nirvana means extinguishing the Three Fires of ignorance, greed, and hatred.
4. *The Fourth Noble Truth of the Elimination of Suffering*: The path for the elimination of suffering (Magga) is a way of life embodying the Eightfold Path or Middle Way.
    a. Right Understanding (Samma Ditthi): Accept Buddhist teaching and practice for your understanding.
    b. Right Intention (Samma Sankappa)
    c. Right Speech (Samma Vaca)
    d. Right Action (Samma Kammanta)
    e. Right Livelihood (Samma Ajiva)
    f. Right Effort (Samma Vayama)
    g. Right Mindfulness (Samma Sati)
    h. Right Concentration (Samma Samadhi)

The Noble Eightfold Path: The Noble Eightfold Path or Middle Way can be divided into three groups: *Wisdom* (right understanding and intention), *Ethical Conduct (Sila)* (right speech, action, and livelihood), and *Mental Discipline (Meditation)* (right effort, mindfulness, and concentration).

Buddha's philosophy is a practical spiritual approach to human life, and therefore, it led many communities to live peacefully.

The Five Precepts: There are five moral precepts for the Buddhist code of living, and they are refraining from:

1. harming living things; most Buddhists are vegetarian, and monks are permitted to meat;
2. taking what is not given;
3. sexual misconduct;
4. lying or gossip; and
5. partaking of intoxicants, such as drugs or drinks.

Lay Buddhists are expected to provide the monks with food, clothing, and all other necessities. Buddhist monks live by ten moral precepts. The ten teachings are the five lay Buddhist precepts, including the five following rules to refrain from:

1. Eating substantial food after midday, from noon to dawn
2. Dancing, singing, and music
3. Use of garlands, perfumes, and personal adornment and jewelry
4. Use of luxurious beds and seats
5. Accepting and holding money, gold, or silver

The Five Skandhas: The Buddha often spoke of the five skandhas (Sanskrit: aggregates) that empty of soul or permanent essence of life. The five skandhas make up the whole of an individual's mental and physical existence.

1. The First Skandha is the physical form of the body *(rupa)*.
2. The Second Skandha is the sensations or human's physical and mental contact with the outside world *(vedana)*.
3. The Third Skandha is the perception of sense objects *(samjna)*.
4. The Fourth Skandha is mental formation *(samskara)*.
5. The Fifth Skandha is awareness or consciousness related to the external world *(samskara)*.

All beings are subject to constant change, as the elements of consciousness are never the same.

The Three Jewels: The simplest way to summarize the Buddhist beliefs is the brief creedal formula of the *Three Jewels*.

1. I take refuge in the Buddha: Refers to historical Buddha as a teacher and enlightened being who inspires and guides us.
2. I take shelter in the dharma: The Buddha's teaching.
3. I take refuge in the Sangha: Refers to the community of monastics and arhats in whom practitioners take refuge

The Middle Way: After Buddha left the palace for the search of knowledge and truth, he examined both extreme self-gratification and extreme self-mortification. At the conclusion, the Buddha defined the Middle Way, which is accessible in everyone's experience. (See Eightfold Path under dharma.)

Scriptures: For over two hundred years, Buddha and the Buddhist monastic community were overshadowed by the powerful Orthodox Brahmanism, and Buddha died as a Hindu. Therefore the Buddha's teachings, which were meaningful and short, were only memorized and recited by his followers; in the meantime, Buddhist teachings were expanding in each Buddhist Council. The Poli Canon or Tripatika was written down in the first century CE, at the Fourth Buddhist Council, under the leadership of Emperor Kanishka. At that time, Buddhism was the largest religion in the world, with many cultures and languages. Also, Buddhists were split into three significant denominations, with many texts and scriptures. In short, there are three significant scriptures:

1. The Theravada (Wisdom of the Elders) Buddhist scripture is the Tripitaka, also called Pali Canon, which includes Buddha's teachings and that of some of his disciples.
2. The Mahayana (the Great Vehicle) Buddhist scripture is the Mahayana Sutras as the Buddha's original teaching.
3. The Vajrayana is a subset of Mahayana. Also, the Theravada Buddhists view Buddha as a teacher, and the Mahayana Buddhists view Buddha as a supreme leader.

Karma: Karma (Sanskrit: action, doing) is defined as a natural law. Any intentional action has a consequence in this life and the future life, and it is karma that leads to birth and rebirth. Evil work, like stealing, injuring, and killing, are bad karmas, and will end up with many births and rebirths; this cycle of birth and rebirth is known as *samsara*. This cycle will bring suffering. Good action, on the other hand, like compassion, generosity, kindness, and helping, will end samsara, and the being reaches *nirvana*, and pain and suffering will end.

Nirvana: Nirvana (Sanskrit: becoming extinguished or blowing out, enlightenment) means the extinction of greed, ignorance, and hatred. Buddha delivered his first sermon and preached the Four Noble Truths, the third of which is a cessation of sufferings (Nirodha). Buddha thought that to overcome desire is to liberate oneself from attachments, and attaining nirvana means extinguishing the Three Fires of ignorance, greed, and hatred.

The Soul: Anatta (Pali: nonself), anatman (Sanskrit: nonself) is the doctrine of Buddha, that there is no permanent substance in humans that we can call the eternal soul or permanent self. Buddha believed and analyzed that the human is compounded of five forces: matter, sensation, perception, mental formation, and consciousness, which are always changing. The concept of nonself is a separation from the Hindu belief in atman (the self). Buddha never found in his analysis that there is an eternal soul. Buddha taught that there is constantly changing consciousness in each incarnation based on the cycle of life, samsara. In today's modern scientific world, the scientists' findings agree with the Buddha's teaching of anatta or no-soul.[93]

Creation: Buddhism is a nontheistic spiritual tradition and has no creator god to describe the origins of the cosmos. Whenever the disciples asked the Buddha about the source of the universe, he refused to answer. He was saying, answering on such things would not lead to nirvana or liberation from dukkha or suffering. One myth told by the historical Buddha in the Aggana Sutta, which is the twenty-seventh Sutta of the Digha Nikaya collection, explains how humans became bound to birth and rebirth (samsara).

Death and afterlife: Buddhists understand that death is a natural part of birth and rebirth (samsara) and may go to incarnation, in which the person's spirit is waiting close by and looking for another body and another life, or may attain nirvana. After the Buddha's death, he was cremated, as was tradition; therefore, many Buddhists have chosen cremation, a testament to the impermanence of life. The funeral preparations are postponed because Buddhists believe that several stages of life, called antarabhava (Sanskrit: intermediary existence), continue a couple of hours after death. Dying Buddhists (or their family) request the presence of a monk or nun for chanting mantras around the dead body.

Buddhists believe in an afterlife, and the reality of the afterlife is divided into three realms (Sanskrit: trailokya):

1. The Realm of Desire (Sanskrit: kamadhatu). This includes hell through the dwelling of animals and humans, a various paradise for good people.
2. The Realm of Form (Sanskrit: rupadhatu). Those reborn into this reality are not subject to the extremes of pleasure and pain. They continue the process of spiritual refinement and education. The realm form beings do not have a sexual distinction.
3. The Non-Form Realm (Sanskrit: arupadhatu). The realm of non-form beings doesn't have desire and sensation, and this realm is the final stage before the state of nirvana.

Rituals and Customs

Buddhists can worship in their house or at a temple. Buddhists pray in their homes. They assign a room or part of a place as a shrine that includes a statue of Buddha and some incense to worship. Buddhist ritual and worship are different from one denomination to the other. Regardless of faith and country of origin, they are honoring Buddha and his teaching. Buddhism understands that it is through the practice of meditation, wisdom, and development of morality one can achieve enlightenment,.

Protective rites: Buddhists are aware of danger and evil influence, and therefore, they include within their stock of religious practices some rituals to protect their beliefs against an evil force. Theravada Buddhist rituals are associated with texts called Partitas, or protection; most of them are assigned directly to the Buddha.

Ordination: Admission to Sangha includes two distinct acts: Pabbajja, or lower ordination, and Upasampada, or higher ordination, are the official declarations as a monk. The process is as follows:

1. Become a Novice First: A man or a woman who intends to take ordination.
2. The Ten Precepts: These are training rules to be observed by novices.
3. In the Olden Days: Besides a secular education, the boys must learn how to read and write.
4. Sacrifice for Good Luck: Before ordainment, shave all hair on the head.
5. The Ordainment Ceremony: Before becoming a monk or nun, he or she must become homeless, and know the 227 precepts.

Meditation: Meditation is a Hindu tradition and ritual. In Buddhism, meditation in Pali is called bhavana, which means to develop. Buddha was born in this tradition, and he adapted meditation because meditation helps to calm and clear the mind, reduces suffering, and ultimately helps attain enlightenment. Buddha taught many different types of meditation.

1. Mindfulness of breathing (Anapanasati): This principle of concentration brings back a person to the present time by focusing on the counting of breath, which helps the person relax.
2. Loving-kindness (Metta Bhavana): This cultivates love and kindness in a person's heart, to connect deeper with all of life and live without conflict with others.
3. Zen meditation: Zen meditation is practiced in East Asia and provides tools to help cope with depression and anxiety. The primary purpose is spirituality.

Prayer: Buddhists don't pray in the hope of favor or help; they pray to connect with compassion, to discover awakening, and to understand the self.

Buddhist ritual prayers are different from one denomination to another. Regardless of faith and country of origin, they are honoring Buddha and his teaching. Buddhists pray at home, work, or temple, and they recite the three jewels: I take shelter in the dharma, I take refuge in the Buddha, and I take refuge in the Sangha. Buddhists also use mantras as a short prayer, which helps them to relax their minds for meditation. The repetition of certain mantra during meditation induces a trancelike state in the person, leads them to a higher level of spiritual awareness. The most effective and widely used mantra is "Om mani padme hum."

Offering and alms: In the Buddha's time, the Buddhists did not have monasteries. The monks were homeless beggars, panning for their food. The only possessions they had were robes and a begging bowl. Later, Buddhists built monasteries where monks and nuns could dwell. Monks leave the monasteries in the early morning, and they walk in single file based on age, the oldest first. Buddhists lay waiting for them, and place food, incense sticks, or flowers in their bowls. Monks and nuns are poor. Therefore, the lay Buddhist's responsibility is to support them with food, clothing, and other needs.

Marriage rituals: Marriage has never been a vital rite of passage in Buddhism. First is the matchmaking process in a Buddhist marriage; a suitable partner is chosen by the parents of the bride- or bridegroom-to-be. The second step is that a friend of the groom's family approaches the girl's family. He carries a bottle of wine and a white scarf. If the girl's family accepts the gift, then it starts with an engagement ceremony in the presence of a monk. Later they schedule the wedding ritual, which could be a small or big wedding. After the wedding, the couple leaves the girl's parental home on a date determined by the couple and the families.

Birth rituals: Buddhism views births, marriages, and deaths as secular rather than religious events, and therefore, no specific ceremonies have been designed for them. Services can be performed based on country and culture, not religion. Within a month, the parents take their baby to a temple and put the baby in front of Buddha's statute. They ask for the blessing of the Buddha, the dharma, and the Sangha.

Summary

Buddhism emerged as a nontheistic religion, and there is no belief in a personal god or goddess, and also as a reaction against the teaching of Orthodox Brahmanism, which was based on a caste structure. The Indian subcontinent religions have the same culture. They believe in karma, rebirth, reincarnation, meditation, and they respect nature. They also have common beliefs, like history, a tradition of asceticism, and literary heritage.

Gotama Buddha was born to a life of privilege into a royal family of the Kshatriya (warrior caste) in Lumbini, in present-day Nepal, in 563 BC. His father was King Suddhodana, the leader and king of the Shakya clan. Gotama Buddha tried to find a cure for diseases in the country where people were dying from leprosy and other acute illnesses, as well as poverty. He did not call himself a leader, a prophet, or a man who received a revelation to bring ordinary Hindus under control. He did not mean to create faith or religion. Later, some of his followers changed the real way of Buddha to a system of belief. They elevated Gotama Buddha as a god.

Buddha's way of life and meditation has considerable influence and practice in the Western world today.

Buddhism emerged from Hinduism, and its two primary components are fear-based.

1. *Karma* (Sanskrit: action, doing) is defined as a natural law. Any intentional action has a consequence in this life and the future life, and it is karma that leads to birth and rebirth or samsara. Evil work, like stealing, injuring, and killing, are bad karmas.
2. *Nirvana* (Sanskrit: becoming extinguished or blowing out, enlightenment) means the extinction of greed, ignorance, and hatred.

CHAPTER 9

Confucianism

Confucius the great teacher and leader, located in Jainshui Confucius Temple, Yunnan, China (Copyright iStock by Getty Images/credit: Aphotostory and Allexandar).

Confucianism is a system of philosophical, social, and ethical practices and beliefs, rather than a religion, founded by Confucius in the sixth to fifth centuries BC. Confucianism doesn't concern itself about the existence of God. It is mostly humanistic, rationalistic, and agnostic; it doesn't have liturgy and theology. Education to attain status and wealth is a primary focus of Confucianism. Confucianism was adopted as an official state religion and philosophical school of China by Emperor Wu-ti of the Han dynasty in the second century BC.

The Chinese people followed Confucianism for more than 2,500 years. It was further developed mainly by Mencius (372–289 BC) and later by Xun Zi. Confucianism influenced the neighboring countries, like Vietnam, Japan, and Korea, on a grand scale, and other East Asian countries on s spiritual, government, political, society, education, and family life level.

Brief History of Religion in Ancient China

Religious practices in Neolithic ancient China go back to the fifth millennium BC. Archaeological evidence at the site of Banpo Village in modern Shaanxi province of 250 tombs, with grave goods, shows a special ritual and care regarding burial practices. The tombs' orientation is west to east, to symbolize death and rebirth. Grave goods show the status of the people in the village. The culture was matrilineal, which means females were dominant.

Products found in the graves of religious figures were women-based goods. There was no evidence of a high-ranking male.

The gods, spirits, and ancestors, like the rest of the ancient world, grew out of people's observance of natural phenomena. These phenomena sometimes scared them or made them happy; as time passed, these beliefs turned into gods and deities, and these deities received names and personalities, and ultimately, rituals were developed to honor them.

Archaeological evidence of the Bronze Age (2200–1600 BC) in the site of Lajia Village in modern Qinghai province and other places shows the proof of religious practices. It appears that the culture could have been changed to patriarchal, which means male dominance. There is evidence of a system of religious changes, in which the people worshipped a supreme god who was king of many lesser deities.

Historical evidence confirms that during the Bronze Age, China's earliest dynasty, the Shang, ruled from 1600 BC to 1046 BC. Religion during the Shang dynasty was polytheistic and animistic. People were worshipping many gods, and ancestor worship was especially vital during the Shang dynasty. Also during Shang, advanced organized religion, a writing system, calendar use, craftsmanship in bronze and ceramic, military technology, and horse-drawn chariots were developed. During the Shang dynasty, there were more than two hundred gods in a Chinese pantheon. Among them was Shangti, the most celebrated ancestor and deity, who controlled the weather, harvest, victory in battle, and the fate of the capital. Shangti was the supreme deity during the Shang dynasty. The Zhou dynasty overtook the Shang dynasty in 1046 BC.

During the Zhou dynasty (1046–256 BC), a new concept of religion, the Mandate of Heaven, was developed. This concept was what Zhou Wu used to overthrow the Shang dynasty. The Mandate of Heaven was believed to be determined by the personal behavior of the ruler. The Mandate of Heaven was expected to possess yi (righteousness) and ren (benevolence). Otherwise, he would be overthrown and leadership would be passed on to another. The gods were thought to watch over the people and would pay special attention to the emperor. The Chinese wore charms and amulets, including those of their ancestors, for protection or blessing, as did the emperor. The state was always in control of religion and never struggled with religious leaders for power—the ruler governed and embedded religiosity into the people.

Confucius

The culture and beliefs of society often change over time because of the influence of individuals and leaders whose intentions can vary from good to evil. Confucius was one of the leaders who was honest and had a good plan for Chinese society; he didn't talk about the two hundred Chinese gods, or the afterlife, or the end of time. One of his disciples asked him about the afterlife. Confucius responded, "I don't know anything about our present life, how I could answer about the afterlife?"[94]

Confucius (Chinese: Kongzi or Kung-fu-Tzu) or Master Kong, was born into the Kong Qiu family; his literary name was Zhongni. Confucius is a Latinized name of the Chinese philosopher who was later introduced to Europe by the Italian Jesuit Matteo Ricci. Confucius was born in 551 BC, in the city of Qufu, in the province of Lu, modern-day Shandong, and died in 479 BC in the city of Qufu. Confucius is the most prominent Chinese teacher, thinker, social philosopher, and political theorist whose teaching in philosophy and politics has influenced the civilization of East Asia.

Confucius's father, She Liang He, died when he was three years old, and his mother raised him under challenging conditions. Confucius first was instructed by his mother; in his teens, he proved an excellent learner. It isn't known who his teacher was, but he found the right teacher to teach, among other skills, music and ritual. He was a master of six arts: music, archery, ceremony, charioteering, arithmetic, and calligraphy. He learned history, poetry, and classical tradition extensively.

Confucius was nineteen years old when he married and had a daughter and a son. His marriage wasn't a happy one. First, his wife died at a young age, and later, his son died one year before Confucius's death.

He began to work at the age of twenty-two for several years for the state. He left the state work and spent many years teaching privately and was hoping to win converts to his political, philosophical, and ethical views. In his late forties and early fifties, Confucius finally had the opportunity to put his ideas into practice. He was appointed chief magistrate of the city of Chung-tu. He made a lot of improvements in the behavior of the people. After one year, he became the minister of crime for the state. His theories worked very well, and almost all crime was eliminated from the province.

The powerholders alienated themselves from Confucius because of his loyalty to the king. Confucius's political career was short-lived when he decided to resign and went into self-imposed exile of almost thirteen years to other provinces, vainly hoping that some other leaders would give him a chance to undertake reforms. He returned home at the age of sixty-seven to spend the rest of his life preserving the classical Chinese traditions.

Confucius's teachings and service for the state were not appreciated during his lifetime. Many years after his death, people realized the strength of his vision, and he got recognition.

The followings are some quotes of Confucius from *Analects* (Lunyu), one of four books written by Confucius and his disciples.

> What you do not want for yourself, do not do to others.
>
> Since you desire standing then help others achieve it since you want success then help others attain it.
>
> It's unimportant how slowly you go as much as you do not stop.
>
> The strength of a nation derives from the integrity of the home.
>
> Courage, wisdom, and compassion are the three universally recognized moral qualities of men.
>
> Our greatest glory is not in never falling, but in rising every time we fall.
>
> The desire to succeed, the will to win, the urge to reach your full potential…these are the keys that will unlock the door to personal excellence.
>
> Silence is a true friend who never betrays.
>
> Everything has a goddess, but not everyone sees it.
>
> Life is straightforward, but we insist on making it complicated.

These lines are the essence of Confucius's vision, virtue, benevolence, and philosophy of universal harmony.

Confucius emphasized the Five Relationships that every individual is part of, and if every person follows these relationships faithfully, society would live in harmony and peace.

1. Emperor and subjects: The emperor would be benevolent to his subjects, and his subjects would be loyal.
2. Parents and children: The parents would love their children, and the children would be obedient (most important).
3. Husband and wife: The husband would be kind to his wife, and the wife would be respectful to her husband.
4. Siblings: The older siblings would be kind to younger siblings, and the younger siblings would be respectful to the older sibling.
5. Friends: Friends were considered equals, and the younger should respect the older.

In 479 BC, Confucius died at the age of seventy-three. Historians recorded that seventy-two of his students mastered the six arts and had three thousand followers.

Leaders

Confucius was the founder of Confucianism, and his thoughts were further developed and compiled by Mencius and Xunzi. Later, in the eleventh century CE, philosopher Zhu Xi revived the Confucian tradition, called neo-Confucianism, and it continued until 1905. There are also some contemporary Confucian advocates who will not be discussed as they are outside the scope of this book.

Mencius (Mengzi, original name Meng Ko), was born in 371 BC, in the ancient state of Zou (Tsou) south of Lu, the home state of Confucius, and died in 289 BC. Mencius has studied in Zisi's school of Confucianism, who was the grandson of Confucius. The development of Orthodox Confucianism from this school earned him the title of second sage, and the Chinese viewed him as the cofounder of Confucianism. Mencius's teachings were preserved by his disciples in a book titled *Mencius*.

Xunzi's original name was Xun Kuang. He was commonly referred to as Xunzi, which means Master Xun, and with an honorary title of Xun Qing. Xunzi lived in the third century BC; his exact birth and death dates and career are uncertain, but he was from the state of Zhao, the modern state of Shanxi. He moved to the state of Chu, the current state of Hubei, where he served as magistrate of a small district in 255 BC. His major work is known as Xunzi; this book comprises thirty-wo well-organized essays written by himself. His doctrine was slightly opposite of Mencius, who proclaimed the innate goodness of humanity. At the same time, Xunzi declared that the nature of human is evil and requires education and religious rituals to cultivate kindness. Xunzi's thought was outside of Confucius's orthodoxy. He gave a minimal role to heaven as a source of morality in comparison to other philosophers of the time. Still, his intellectual approach described him as the molder of ancient Confucianism.

Dong Zhongshu, also known as Tung Chung-Shu, scholar and philosopher, was born in 179 BC in the village of Guangchuan in the province of Anhui, China, and died in 104 BC. Dong Zhongshu merged Confucius and the school of Yinyang (ancient Chinese divination practiced as early as the fourteenth century BC). He was respected as a worthy scholar by Emperor Wu (140–87 BC) of the Han dynasty, who appointed him as his chief minister. Dong Zhongshu dismissed all the non-Confucians from the government. Dong Zhongshu established Confucianism as a state system of religion in 136 BC, and it was the foundation of the political philosophy of China for two thousand2,000 years. His major work was the interpretation of the Confucian Classic Spring and Autumn Annals.

Zhu Xi (Chu His; also called Zhuzi or Zhufuzi), was born October 18, 1130, in Youxi County, Fujian Province, China, and died on April 23, 1200. Neo-Confucianism started in the middle of the ninth century for the revival of Confucian political culture and philosophy. At the same time, Buddhism and Daoism had both developed into powerful forces in Chinese culture and life. Neo-Confucianism reached its new levels of intellectual, social, and political innovation at the beginning of the eleventh century under the leadership of the greatest philosopher, Zhu Xi, in agreement with

Mencius on the innate goodness of humankind, restoring Confucian ideas to prominence. By the fourteenth century CE, Zhu Xi's version of Confucian thought, known as the teaching of the way, became the standard curriculum for the imperial civil service examination.

Scriptures

Confucius didn't believe in God; he was a teacher, thinker, social philosopher, and political theorist whose teaching in philosophy and politics altered the ancient Chinese political and religious system. Confucianism has two primary scriptures: The Five Classics (Chinese: Wujing) which, according to legend, was compiled by Confucius, even though the significant parts dated after Confucius's death, during the Han dynasty (206–220 CE). The other scripture is The Four Books (Chinese: Sishu), among the four, *Analects* or conversation (Chinese: Lunyu) contains direct quotations from Confucius as recorded by his disciples.

The Five Classics (Wujing) Confucius compiled was the first step for altering the ancient Chinese political and religious system.

1. *The Classic of Changes or I Ching (Chinese: Yijing)*: This contains all the systems of divinations and many divinations practiced at present.
2. *The Classic of History (Chinese: Shujing)*: This contains the history of the Xia or Hsia (2070–1600 BC), Shang (1600–1046 BC), and Zhou or Chou (1046–256 BC) dynasties. This the first narrative of ancient China.
3. *The Classic of Poetry (Chinese: Shijing)*: This contains 305 poems for different purposes: love and marriage, daily lives, agricultural concerns, and war; also, songs and hymns. Confucius selected the 305 poems from a much broader collection.
4. *The Collection of Ritual (Chinese: Liji)*: This contains governmental organization, social standards, and ritual behavior. This document maintains proper ritual behavior and harmony in the empire.

5. *The Spring and Autumn Annals (Chinese: Chunqiu)*: This contains the chronological record of significant events of the state of Lu, the modern province of Shandong.

The Four Books (Chinese: Sishu or Ssu-shu) were the official resource for civil service examination in China from 1313 CE to 1905 CE, and Confucius and his followers wrote them as the core teaching of his philosophy. The Sishu was published as a unit in 1190 CE, and was published in four parts without logical order.

1. *The Great Learning (Chinese: Daxue or Ta-Hsueh)*: This contains a guide for moral self-cultivation and a detailed text of the ethics and politics of a good ruler. It teaches the rulers how to learn and practice a Great Way, how to rule the whole world. It's believed that the Great Learning was completed by Confucius disciple Zingzi (Zeng Shen) from 505 to 435 BC. Zhu Xi suggested the Great Learning must be the first to be read among the Confucian classics.
2. *The Doctrine of the Mean (Chinese: Zhongyong)*: This discusses things like the Way of Heaven, religious sacrifices, motion, and spiritual beings. This book is believed to have been written by Zisi, the grandson of Confucius, and teaches how to maintain ideal balance and harmony in one's life. The precept of the Mean and the Great Learning are both excerpts from The Collection of Ritual (Liji), one of the Five Classics.
3. *Analects or conversation (Chinese: Lunyu)*: This contains direct quotations, teaching, and dialogue of Confucius with his disciples as recorded by his disciples. The *Analects* also draws attention to the importance of filial piety, ancestors, good governance, virtue, and ritual.
4. *The Book of Mencius (Mengzi)*: This book is the most extended text of The Four Books and contains the teachings of Mencius, who believed in the goodness of human nature. It means that humans are innately ethical and moral, but we must learn how to nurture and cultivate those seeds. Mencius was regarded as the second sage in Confucianism, next to Confucius himself.

Denominations

Confucianism is not a religious system; the founder, Confucius, did not discuss or concern himself with the existence of gods or goddesses. He studied ancient Chinese history and compiled the Five Classics (Chinese: Wujing). It was the first step for altering the ancient Chinese political, social, and religious system. Confucius revered dead ancestors and supported filial piety to help create a harmonious society and a peaceful environment. After Confucius's death, Confucianism split into eight schools: (1) Mencius, (2) Xunzi, (3) Dong Zhongshu, (4) Song Confucianism, (5) Ming Confucianism, (6) Korean Confucianism, (7) Qing Confucianism, and (8) Modern Confucianism. Only a few will be discussed here.

Mencius: Mencius's doctrine was described in two books. The first book is the Doctrine of the Mean (Chinese: Zhongyong), which was edited by the grandson of Confucius, and the second book is the Book of Mencius. Mencius's school based his teaching in Confucius's doctrines, while forming its theories on human self-cultivation and self-transformation. The core of Mencius's teaching is the belief that human beings are born with the knowledge of the good and the ability to do good. All humans are born with what Mencius described as benevolence, righteousness, respect, and the capacity to distinguish right from wrong. Mencius's teaching lasts to this day.

Xunzi: Xunzi was the last thinker of the Confucian classical period, for whom a school of thought is named. Xunzi believed, like Confucius and Mencius, that education and self-cultivation are essential. He emphasized the need for authority, intense literary work, and social structure to eliminate the evilness of human nature. Xunzi rejected the doctrine of the Mandate of Heaven proposed by Confucius and Mencius, and he said heaven does not care about human affairs; it is a natural world. Xunzi significantly stressed formal rituals and music to stop a disordered society.

Zhu Xi: Neo-Confucianism had to create several metaphysical systems, which was not in Confucius's and Mencius's books, to regain the Confucian intellectual dominance, which they had lost to Daoist and Buddhist philosophy. Neo-Confucianism, or revival of

Confucian thinking, emerged in the early Song dynasty (960–1279 CE). It reached its new levels of intellectual, social, and political innovation at the beginning of the eleventh century under the leadership of the great philosopher Zhu Xi.

Beliefs

Confucianism doesn't have God; it's a socio-philosophical movement to create a better society and emphasizes harmony within human relationships. Confucius didn't concern himself about the afterlife, soul, death, or creation.

Ren and li: Ren can be translated from Chinese as humanity, humanness, goodness, and benevolence, which is the primary virtue of Confucianism. Every human is capable of exhibiting ren to promote a flourishing human society. Confucius expressed, "Whenever I want ren, it is as close as the palm of my hand."[95]

Li is a complement virtue to ren; it can be translated from Chinese as ritual, proper conduct, propriety, or etiquette. The concept is part of a ritualized ancestor in a canonical form; Confucius modified it to apply to all activities in life.

Tian: Tian means Heaven or sky. He is the supreme being of lesser gods in aboriginal Chinese religion, and is an integral part of Confucianist philosophy and is associated with the highest deity as well as the domain of ancestors. Tian, during the Shang and Zhou dynasty periods, was offered sacrifices as the highest deity.

Tianming: Tianming (Tian Ming) means Mandate of Heaven. In Chinese Confucius doctrine that Heaven is awarded directly upon the emperor, the son of Heaven or Tianzi, the right to rule. This doctrine started in the early Zhou dynasty. It requires the emperor to possess yi, or righteousness, and ren, or benevolence. Confucianism teaching expresses that in the case of anarchy, poverty, or natural disaster, the emperor had not only lost his right to rule, but also, he should be removed by revolution if necessary.

Ancestor worship: Ancestor worship is very likely the continuation of primitive religion in China. It goes back to the Neolithic period (6000 BC); it's similar to the primitive religion of Andaman

Islanders. The family was the essential element of Chinese society. This culture is maintained by Confucianism; the twin pillars of filial piety and one's dead ancestor to help create a harmonious society and peaceful environment. The ancestor worship is not like god worship in other religions; it is just reverence to the ancestors.

Rituals and Customs

Ancient Chinese worshipped more than two hundred gods and ancestors. This culture became an integral part of Chinese rulers beginning with the Shang dynasty (1600–1045 BC) and the Zhou dynasty (1046–256 BC). Confucius, although he didn't believe in God, he was part of the culture, and he revered dead ancestors and supported filial piety to help create a harmonious society and a peaceful environment with many rituals.

Funeral rituals: The funeral ritual is a significant part of Chinese social life. They believe the dead continue to influence the life of the living. Each member of the family wears a specific colored mourning garment, depending on the relation to the deceased—the mourners wear a mourning pin after the funeral from forty-nine to one hundred days. The corpse will be washed ritually by the oldest son with water, then scented and dressed in a couple of layers. A pear will be placed over the mouth of the deceased to make the journey easy through hell. Musicians lead the funeral procession to frighten the malicious spirits away. After the burial, the family watches the burial site. The ancestor's name is carved on a plate and placed in the ancestral shrine of the family hall as part of the dead ancestors.

Marriage rituals: Confucians believe marriage is a crucial part of society and perpetuates the extended family. Marriage is arranged by the parents, and children have no say in the choice of husband or wife. A dowry from the groom's parents will be negotiated with the bride's family and, after the agreement, will be paid to the bride's parents. After the wedding, the new couple becomes part of the groom's family, and the new couple does not own property. The marriage based on Confucianism is an entrance to the political stability and peace of society.

Birth: Confucians believe a newborn is a link in the chain of existence between the past and the future generation. The Chinese prefer to have a son, and the pregnant mother, regardless of boy or girl, is very much valued. The spirit of the fetus protects the mother during pregnancy and delivery. After the birth of a son, at the end of the first month, the family will give a party to the extended family and friends to announce the arrival of the new member of the family.

Ancestral shrines: An ancestral shrine, hall, or temple, also called a lineage temple, can be in people's houses for their own family or a structure anywhere. Ancestral temples are part of the Confucian culture—the central feature of the ancestral shrines in the ancestral tablets that incorporate the ancestral spirits—the plates placed in order of seniority. Statues can represent the gods and ancestors. The ancestral churches or temples can also be used for other purposes, like festivals, collective rituals, community-related events, and weddings.

Festivals and Holidays

Confucius's birthday: Confucius's birthday is on the twenty-seventh day of the eighth month of the Chinese calendar. It is a yearly festival to pay respect to the founder of Confucianism and lasts several days. Chinese Teacher's Day is celebrated a few days earlier.

Chinese New Year: Confucians celebrate Chinese New Year on January 31.

Tomb-Sweeping Day or Qing Ming: Qing Ming is another festival, and is celebrated on the fourth or fifth of April, fifteen days after the spring solstice. On this day, people go to their ancestors' graveyard to clean and remove the weeds. Significant observances take place at the Confucius Temple in Qufu, province of Shandong, where Confucius was born and died. On this day, people play music, dance, make sacrifices, and much more.

Summary

Confucianism is a system of social, philosophical, and ethical practices and beliefs rather than a religion. It's a socio-philosophical

Confucianism

movement to create a better society, emphasizing harmony within human relationships. Confucianism was founded by Confucius in the sixth to fifth centuries BC, and Chinese people have followed his movement for more than 2,500 years. Also, this system of socio-philosophical action influenced all the neighboring countries. Confucianism is a nontheistic system that does not have gods or goddesses.

Confucius (Chinese: Kongzi or Kung-fu-Tzu) or Master Kong was born in 551 BC and died in 479 BC at the age of seventy-three. Confucius is the most prominent Chinese teacher, thinker, social philosopher, and political theorist who altered the ancient Chinese political and religious system. His teaching in philosophy and politics influenced the civilization of East Asia. Confucius wrote and compiled the Five Classics or Wujing, which changed the landscape of the Chinese political and religious system. Confucius didn't concern himself about the afterlife, soul, death, or creation. According to the legend, one of his disciples asked Confucius about life after death. He humorously responded, "Why should we concern and understand the afterlife, while we do not fully understand this life we are in now?"

Confucianism is not fear-based because Confucianism is a socio-philosophical system to create a better society and emphasizes harmony within human relationships. Confucius stressed five relationships.

1. Emperor and subject: The emperor would be benevolent to his subjects, and his subjects would be loyal.
2. Parents and children: The parents would love their children, and the children would be obedient (most important).
3. Husband and wife: The husband would be kind to his wife, and the wife would be respectful to her husband.
4. Siblings: The older siblings would be kind to younger siblings, and the younger siblings would be respectful to the older siblings.
5. Friends: Friends are considered equals, and the younger should respect the older.

CHAPTER 10

Taoism

The largest Song dynasty statue of Laozi, the founder of Taoism (Copyright iStock by Getty Images/credit: Ryan Whyte and Eloku).

Taoism (the Wade-Giles system of the 1900s), also spelled Daoism (Hanyu Pinyin system of the 1950s), is an aboriginal religion/philosophy attributed to Lao Tzu and began with the writing of the Tao-Te-Ching, or the Book of the Way, in 300 BC. Taoism is a system of folk religion and philosophy with no real founder, as well as no date of founding.

The Tao-Te-Ching is the beginning of Taoism; who wrote it first is not known. The teaching of the Tao-Te-Ching came to fruition by thinkers like philosopher Yang Zhu (also spelled Yang Chu), who was born in 440 BC and died in 360 BC. He was an advocate of naturalism. Philosopher Zhuangzi (also spelled Chuang-Tzu), was born in 369 BC and died in 286 BC. He contributed significantly to the interpretation of Taoism.

Taoist ethics emphasize the fundamental principle of Wu Wei; *Wu* means nothing, and *Wei* means action or as a whole nonaction. This principle says that Taoists should live ponderously in compliance with the natural flow of the cosmos (Tao) and with Three Treasures: compassion, humility, and frugality.

Scholarly research advocates that Taoism as a religious tradition started around the second century CE, with a renowned teacher Chang Tao-ling (also spelled Zhang Daoling), who was born in 34 CE and died in 156 CE. Zhang Daoling declared he dreamed of a revelation from Lao Tzu (sage Laozi) to spread the teaching of the Tao, and he composed the Xiang'er commentary on the Tao-Te-Ching.

He founded a movement called the Way of the Celestial Masters, known as the Way of the Five Bushels of Rice (Chinese: Wudoumi). This movement created the first and most influential branch of religion in Taoism, and later, many other sects and schools were developed.

Taoism does not recognize any other omnipotent god other than the cosmos, which springs from the Tao, and the Tao impersonally guides us on our way. Lao Tzu is not a god, but he is revered as the god of Taoism and personification of the Tao.

Taoism believes in many minor gods, who are the gods of specific tasks. Taoism is the same as Confucianism as a continuation of ancestral religion and is related more to shamanism. Taoism shares a strong element of shamanism in its belief, and that is the physical and the spiritual.

Taoism spread into all Asian cultures and especially Japan, Korea, and Vietnam. All religions in China developed from ancient beliefs; see the section "Brief History of Religion in Ancient China" under "Confucianism."

Laozi

Whoever was the leader in this the Taoism movement was living during the warring states, and therefore, he was against big government and unjust laws created by the government. He believed that people misbehave because they are forced to through weak government and unfair rules. It is thought that Laozi or Lao Tzu, whose name means Old Master, never existed; there are many myths about him.

Historian Sima Qian (145 BC–86 BC), tells a myth of Lao Tzu. "He was so frustrated by his inability to change people's behavior, then he decided to go into self-exile. As he was leaving China through the western pass, the gatekeeper Yin Hsi stopped him because he recognized him as a philosopher. Yin Hsi asked Lao Tzu to write a book for him before he left the civilization forever and Lao Tzu agreed, and he sat down on a rock next to Yin Hsi and wrote the Tao-Te-Ching or the Book of the Way, and Lao Tzu continued to his travel and vanished in the mists."[96] Sima Qian adds about his age, "Maybe Lao Tzu has lived one hundred and fifty years, and some say more than two hundred years." It appears that he was a legendary person and never existed.

Lao Tzu was one of the revered figures in all circles in China; people worshipped him as a saint or a god. Confucius recognized him as a revered philosopher, and for Taoists, he is one of the most important divinities.

Leaders

Tao Tzu (Laozi, 601–531 BC) is the legendary founder of Taoism, and is also a legendary author of the Tao-Te-Ching, the best-known Taoist text. Yang Zhu (Yang Chu) and Zhuangzi (Chuang-

Tze) contributed to the philosophical aspect of Taoism, and Chang Tao-ling (Zhang Daoling) contributed to the religiosity of Taoism. There are a few more leaders who contributed to Taoism, like Zhang San-Feng and Ursula K. Le Guin, which will not be discussed in this book.

Yang Zhu (Yang Chu): Born in 440 BC and died in 360 BC, he was the advocate of naturalism. Yang Zhu was the first Chinese philosopher to talk about human nature. Yang Zhu believed that human beings should live pleasurably. He is described as a hedonist, and is alleged to have saved the world by sacrificing one single strand of his hair.[97] He was not that famous except that he wrote the seventh chapter of the Taoist scripture Lieh Tzu. Also, his thought was in a few sections of the philosophical and literary classic the Zhuangzi, written by Zhuangzi (Chuang Tzu).

Zhuangzi (Chuang Tzu): A philosopher whose original name was Zhuang Zhou (Zhuang Chou) was born in 369 BC in Shangqiu, Henan Province, China, and died in 286 BC. He is the best interpreter of Taoism, and he wrote the Zhuangzi texts, the second most famous Taoist text, more comprehensive than the Tao-Te-Ching. Zhuangzi composed thirty-three chapters; he wrote the seven chapters and the inner books. His followers produced the rest of the sections.

Zhuangzi believed life is the ongoing evolution of the Tao, which is that right or evil follows nature. When he was close to death, his followers asked him about his funeral. He answered, "Nature will be my inner and out coffin."[98] His followers responded that crows and buzzards would eat him. He replied, "What difference it makes, above ground are crows and buzzards and undergrounds are ants and worms."

Zhang Daoling (Chang Tao-ling): Zhang Daoling was born in 34 CE in Pei County, Jiangsu Province, and died in 156 CE in Hanzhong. Zhang Daoling studied Taoism and Buddhism, and he trained in the Confucian classics. He liked the teaching of Lao Tzu and the arts of longevity and immortality. He composed the Xiang'er commentary on the Tao-Te-Ching. He founded a movement called the Way of the Celestial Masters, known as the Way of the Five Bushels of Rice (Chinese: Wudoumi). This movement created the first and most influential branch of religion in Taoism, and he expressed the importance of religious organization to his followers.

He endowed Lao Tzu with the title of T'ai-Shang Lao-Chun, which means the Great Lord on High, and he, his son Chang Heng, and his grandson Chang Lu became the cult's leaders. Zhang claimed that he received a revelation from the sage Lao Tzu for teaching, and was also given the power to heal the sick and ward off evil spirits. As a result of teaching, healing the sick, and familiarity with talismanic magic, he became enormously influential and famous. Zhang Daoling participated in the uprising against the Han dynasty and became a major political figure in Hanzhong.

Zhang San-Feng (Chang San-Feng): Zhang San-Feng was born on April 9, 1247 CE, died in 1447 CE. He was a cultural hero throughout China, a legendary Taoist immortal; he lived for two hundred years. When he reached twenty years of age, he left home and wandered around with an ascetic life. After ten years, Zhan San-Feng moved to the Wundong Mountains, where he taught meditation, alchemy, herbalism, and traditional medicine. Zhan San-Feng is the originator of tai chi chuan, a type of martial arts known for deliberate and slow movement, and started to integrate the practice of martial arts with internal alchemy and taught other monks tai chi for their protection.

Scriptures

Taoism includes a vast collection of books, such as a code of conduct, genealogy, revelation, philosophy, sacred diagrams, and many other disciplines.

Tao-Te-Ching: Tao-Te-Ching is at the heart of Taoism; it is a short text of around five thousand Chinese characters in eighty-one short chapters. Tradition has it that these texts were written in the sixth century BC, when Lao Tzu was disappointed with the court life. He decided to go into self-exile through the Western Pass. The gatekeeper asked him to write a book. He agreed, and wrote the book called Tao-Te-Ching, which means the Way and Its Power. Most scholars now believe Lao Tzu is a mythological figure, and several people wrote the book over an extended period. Here is one passage of the text: "The three greatest treasures are simplicity, patience, compassion. Simple in activities and thoughts, you return to the source of being. Compassionate toward yourself; you coexist in harmony all beings in the world. Patients with both enemies and friends, you accord with the way things are."99

Zhuangzi: The second most important book of philosophical Taoism is the Zhuangzi, which was written by fourth century BC philosopher Zhuangzi. His thinking was more concerned with the general public than the rulers. Zhuangzi contemplated the nature of reality and reflected on the infinite transformations that occur in life and death, which he believed was a blending with Tao. It is thought that Zhuangzi was written by many others, except seven chapters of thirty-three chapters are attributed to him.

Huainanzi (Huainanzi or Huai-Nan-Tzu; Chinese: Master Huainan): Great Words from Huainan is another excellent work of philosophical Taoism. This Chinese classic was written in the second century BC, with the assistance of the Huainanzi (Liu An), who was the grandson of Gaozu, the founder of the Han dynasty, a nobleman and prominent Taoist philosopher. The Huainanzi, also known as Huainan Honglie, is composed of twenty-one loosely connected sections on cosmology, metaphysics, matters of state, and conduct.

Liezi (Chongxu Zhide Zhenjing): Lieh-Tzu (Lieh-Tzu or Liezi; Chinese: Master Lie), whose original name was Lie Yukou, was one of the significant Taoist philosophers of the fourth century BC. Lieh-Tzu created the fundamental belief of Taoist philosophy, and he wrote the Taoist work Liezi, also known as Chongxu Zhide Zhenjing. Lieh-Tzu explains the Tao and its changes in a variety of stories showing the miracles found in creation.

Daozang (Daozang, or Canon of the Way, or Tao Tsang; also called the Taoist Canon): This work is designated as an imperially sponsored collection of Taoist texts and is kept in the imperial libraries. An example is the Taoist Canon of the Ming period, which is almost 1,500 texts found in the present Taoist Canon that are formally divided into Three Caverns, or Three Grottoes, or sandong; Four Supplements or Four Auxiliaries, or sifu; and twelve subdivisions called Categories. A total of 1,500 texts run to 1,120 volumes.

Denominations

Taoism does not have formal denominations, as it originally started as a philosophical system, and later became a religious system for the expansion of Taoism in competition with Confucianism and Buddhism. Therefore, there are two divisions of Taoism: philosophical Taoism and religious Taoism.

The philosophical Taoism concerns topics in the Tao-Te-Ching and Zhuangzi, such as emptiness, detachment, spontaneity, receptiveness, understanding of the world, search for a healthy and long life, and a harmonious relationship with the Tao.

The religious Taoism is a continuation of ancient Chinese folk religion and believes in the existence of supernatural beings who enter into relation with human beings, such as gods, ghosts, and ancestral spirits. Gods are generally on the side of righteousness; ghosts are dangerous spirits of the departed, as well as the ancestral spirits, who are conciliated with offerings.

Zhang Daoling was at the forefront of religious Taoism, and he created the religious movement called the Way of the Celestial Masters to expand Taoism under his leadership. Zhang Daoling participated in the uprising against the Han dynasty and became a major political figure in Hanzhong.

Religious Taoism falls into two minor denominations: The Way of Orthodox Unity or Southern Taoism popular in Taiwan and South China, and the Way of Complete Perfections or Northern Taoism, popular in mainland China.

Southern Taoism: Southern Taoism is the successor of Cheng-I Taoism. Its roots are in the school of the Celestial Master, the first founder, Zhang Daoling, who received a revelation from the sage Lao Tzu for teaching, and also was given the power to heal the sick and ward off evil spirits.

In Southern Taoism, priests are married. They perform liturgies like the Chiao, and their practices include several rituals to restore harmony in the world, such as rites that unite the local community with the cosmos, exorcise evil spirits, heal the sick, and use a talisman.

Northern Taoism rejects Southern Taoism's liturgical practices and rituals, and they distance themselves from Southern Taoism. They believe in studying Taoist scriptures to nurture their knowledge and become a better person. They pursue self-preservation practices that date back to the Classical Neiye, or Inward Training, like internal alchemy and the pursuit of immortality. Wang Chung-yang (Wang Che) founded the Way of Complete Perfection, or Northern Taoism, in the twelfth century CE. Its headquarters are at White Cloud Abbey in Beijing.

Beliefs

Originally, Taoism was started as a philosophical system by the Chinese philosopher Lao Tzu, who grew up in ancient China. Later, Zhang Daoling added the religious order to Taoism and gave Lao Tzu the title of T'ai-Shang Lao-Chun, which means the Great Lord on High, and brought him to the level of a god. The fundamental belief of the Taoist is to live in harmony with the Tao, or the Way, ultimately merging with Tao and becoming immortal.

Priests (Chinese: Taoshi): Because of religious inclusion into Taoism, priests play an important role, like in annual festivals, rites of passage, the ritual of purification, exorcism, healing, the ritual of offerings, and maintaining temples.

Deities: Taoists believe the cosmos sprang from Tao, and the Tao nonbeing guides things in the right way. The Taoist pantheon of gods borrowed from various religions and sects, like Buddhism, Hinduism, and China's own folk beliefs. Tao Tzu gradually became the highest deity. First, he was a legendary figure, teacher, and writer; second, his image blended with the Yellow Emperor, and he became a royal confidant. Third, Zhang Daoling promoted him as T'ai-Shang Lao-Chun, meaning the Great Lord on High, brought him to the level of a god, and finally as a creator god. Shortly after that, Tao Tzu joined the Three Pure Ones, similar to the Trikaya (Dharmakaya, Samboghakaya, and Nirmanakaya) of Buddhism. The Three Pure Ones are the following:

1. The Jade Pure One (Yuqing) is honored as the source of learning.

2. The Supreme Pure One (Shangqing) reveals Taoist scriptures to the lesser gods and humans.
3. The Grand Pure One (Taiqing) is the manifestation of Tao Tzu, which is the chief divine person.

They represent three aspects of the divinity innate in every living being.

Creation: Pan-Gu is the leading figure in Taoism legends of creation. Taoists do not have a single creation story like other major religions. They believe the primordial Tao existed before the universe. Taoists think creation is an ongoing process and continually evolves; the whole process is self-generating and self-perpetuating. They also think nature brings order to the universe without fulfilling any divine purpose.

Tao: Tao means the Way, the process of the cosmos by which everything changes, and to follow a life of harmony. Also, Tao can be defined as a system of guidance, which means it is the source and guiding principle of all things that exist and all that happens.

Ethics: Taoist ethics requires the person to live in harmony with all things and people rather than doing kind acts. Taoists and all East Asian religions progress over time, and they have revised behavioral guidelines on ethical considerations. Like every religion or system of government, the Taoist disapproves of stealing, killing, lying, hurting others, an unfaithful spouse, and also promotes kindness, being helpful to others, and ethical behavior.

Wu Wei: *Wu* means no or without, and *Wei* means do or make; Wu Wei means no doing, not making, nonaction, or actionless; action in nonaction. Wu Wei found in Tao-Te-Ching is a central principle in the philosophy of Taoism. This law states that humans should live in agreement with the natural flow of the universe (Tao) for human happiness to occur. Wu Wei brings harmony with one's natural surroundings.

The Three Treasures: The Three Treasures, or the Three Jewels (Chinese: Sanbao), presented in chapter 67 of the Tao-Te-Ching, are the basic ethical guidelines of the Taoist tradition, and show a submissive relationship with the Tao. They are Qi (compassion), Jing (frugality), and Bugan Wei Tianxia Xian (humility). Nobody can embody these virtues perfectly, but can practice them in daily life to the best of their ability.

Scriptures: The main book of Taoism was compiled around 300 BC, and it's called the Tao-Te-Ching. These texts were not written by Lao Tzu, but were written by many people over an extended period. Philosopher Zhuangzi wrote the other short book, self-titled Zhuangzi. Zhuangzi wrote intimate stories that ordinary people could relate to without difficulty; it is a collection of the wisdom of many. Zhuangzi took Tao-Te-Ching's abstract ideas and applied them to everyday situations.

Death and afterlife: Taoism originally started as a philosophy; a couple of hundred years later, Zhang Daoling included religion for more participation and expansion of Taoism. Taoism is not interested in death; it's part of the cosmic cycle. This means we are born out of emptiness and go back to a vacuum, and life and death are two aspects of reality, the unchanging Tao.

Zhuangzi's wife died; soon after her death, he sat cross-legged, pounding on a drum and singing. When his best friend, logician Hui Tzu, expressed condolences, Zhuangzi acknowledged his initial grief, and he explained to his friend that the cycle of the universe brought her to life and took her back to death. Zhuangzi said, "Since life and death are each other's companions, why worry about them? All beings are one."

The Taoist priest has a vital role in the death ritual and burials. The priest communicates with the dead and restores harmony in the family. After the body is buried, the family provides things for the dead to be comfortable, like food, drink, incense, and the priest will determine if the family is suffering from a discontented ancestor. The priest makes sure to perform funerary rituals to settle the soul (hun) into the old tablet in the domestic altar. The period of grief varies from three to six days, depending on the geographic location.

Taoism has no doctrine of the afterlife; Taoists preferably stress longevity, good health through martial arts or tai chi, meditation, and diet. What happens to Taoists after death is not essential. Just to mention, Taoism developed in ancient China, so one cannot ignore respect to ancestral belief, which is the culture and history of the Chinese.

Rituals and Customs

Rituals, a series of acts regularly repeated that make up religious observance, became an essential aspect of Taoist worship from the time of Zhang Daoling, around the second century CE. Taoist rituals include meditation, purification, exorcism, healing, and offerings to deities.

Meditation: Taoist traditional meditation practices include mindfulness, contemplation, concentration, and visualization. Practicing or learning meditation is a huge task in which the meditator has to renounce all sense of self and become one with Tao, like a stream that connects to the river. The basis of meditation can be found in the Tao-Te-Ching, Lao Tzu, verse 16: "Be empty, embrace the calmness of peace. Watch the operation of all creation; perceive how endings become beginnings."

Alchemy: The Taoist practice of alchemy is mainly concerned with transforming humans to have a long life and come closer to Tao. There are two types of alchemies: internal (Neidan) and external (Waidan). The internal alchemy takes place in one's self, not in a real lab, and involves meditation, martial arts like tai chi, and yoga to bring about physical and mental changes. The external alchemy is prepared in outside laboratories by mixing chemicals to produce magical elixirs that would make humans immortal.

Tai Chi: Philosopher Zhuangzi had some new inspirations for the movement philosophy of tai chi in his writings. There is also a long tradition of Taoist monks practicing exercises of tai chi and tai yin or Taoist breathing. At the same time, the precise history and origin of tai chi is obscure, but they are mentioned in the Chinese chronicle around 122 BC. The principle of centeredness, softness, and slowness of Tai Chi reflects the teaching of the Tao-Te-Ching about balance and gracefulness. In around 1459 CE, the purported founder of tai chi chuan, the monk Zhang San-Feng, was honored by Emperor Chengzu with the title of immortal.

Talismans: The Taoist talismans are the continuation of ancient Chinese talismans, and those are objects thought to have the power to bring aid to those who are in need. The Taoist talismans are called

Fu, and they manifest spiritual or psychic energy. The power of each talisman is different based on sect and location—the Tao of cult grants the talismans' code to the people. Talismans can only be written words, symbols, or characters by very qualified people who have undergone long training and are in a state of purity. Talismans are used to invoke deities, become wealthy, curse an enemy, activate good chi, dispel sorcery, and prevent burglary.

Chiao: Chiao (Jiao) is a major community-wide ceremony, a ritual of cosmic renewal that is made up of several small rituals, bringing harmony and order to many layers of the universe. It is regularly celebrated every three to five years. A shortened form of this ritual requires the villagers to present an offering to their gods to bring prosperity to the villagers. The Taoist priest devotes the offering in the name of the donors and asks the deities to bring peace and happiness to the village. The priest and assistants chant, dance, and play instruments.

Ancestors' rites: Ancestors were worshipped since before the Shang dynasty (1600–1046 BC); their most celebrated ancestor and deity was Shang Di (Shang Ti). During the Zhou dynasty (1046–256 BC), they are replaced by Heaven or Tian. Ancestor rites place a high value on family and their continuity. An elaborate ancestor rites ceremony requires an altar with the incense in the center, flowers and candles on each side, and the tablets. In early April, the families celebrate a festival in which they fix and clean their ancestors' tombs.

Birth and infancy: Taoists believe birth is a transition from one form of life, inside the womb, to another form, outside the womb, which is part of cosmic events. Chinese honor a newborn as a link in the chain of existence between the past and the future generation. The Chinese prefer to have a son, and the pregnant mother, regardless of carrying a boy or girl, is very much valued. Tai Shen, or the spirit of the fetus, protects the mother during pregnancy and delivery. After the birth, at the end of the first month, the family will give a party to the extended family and friends to announce the arrival of the new member of the family. During this month, the mother stays home, eating prescribed food and avoiding certain activities.

Marriage rituals: The Taoist wedding can take place anywhere. In a traditional Taoist wedding, the bride and groom face each other while sitting at the center of the Pa Kua or Bagua, meaning eight trigrams. Inside the Pa Kua are the eight forces of nature, with eight candles and divided into a yin and yang. The yin group forces are Wind, Water, Earth, and Lake. The Yang group forces are Heaven, Mountain, Fire, and Thunder. The bride sits protecting an urn filled with water, and the groom sits protecting a jar containing a lit candle. The couple begins the ceremony by lighting all the candles. After reading the spiritual union of Tao, the master of the service greets everyone and reads the understanding of Tao, and the couple exchange vows and rings. The bride's water is poured onto the groom's fire, creating steam, joining together two opposing elements and making one. Traditionally, the couple wears red because red in China symbolizes good luck.

Festivals and Holidays

Many Taoist festivals are ancient traditional Chinese holidays, and most of them are also common in Confucianism and Buddhism. The date of celebration varies based on geography and sects.

Laba Festival: This festival is celebrated on the eighth day of the twelfth month (in the Chinese calendar, this is in January). The Laba Festival marks the day when the Buddha became enlightened, according to tradition.

Chinese New Year: This is the first day in the Chinese calendar of the full moon between January 21 and February 20. At this festival, cleaning the house is very important; the Chinese decorate their homes, and families gather for a traditional dinner.

Lantern Festival: The lantern festival is the first full moon of the first month in the Chinese calendar, between February and March. This day is also the birthday of Tianguan, one of the Taoist gods of excellent fortune. On this day, they eat dumplings called Tangyuan. This food represents happiness and family unity. According to tradition, the Chinese hang red lanterns on the streets and homes for luck and prosperity.

Tomb-Sweeping Day: Tomb-Sweeping Day is celebrated on the fifteenth day after the spring equinox or April 4 or 5. This day is also known as Ancestors' Day or Qingming. This holiday dates back to the Zhou dynasty. The modern revision was made by the Tang dynasty when Emperor Xuanzong decreed the celebration of ancestors would be limited to a single day of the year. Families go to burial places to clean the tombs and leave offerings of tea or food and burn joss paper.

Dragon Boat Festival: The Dragon Boat Festival or Duanwu Festival is celebrated on the fifth day of the fifth month of the Chinese calendar, in May or June. Duanwu has several meanings: a celebration of masculine energy, a time of respect for elders, or a commemoration of the death of the poet Qu Yuan as a prominent member of the Zhou dynasty. He tried to eliminate corruption in the Zhou dynasty. His opinions were not popular, and he was forced out of his position. When the Zhou dynasty fell, he committed suicide by jumping in a river on this day.

Hungry Ghost Festival: This festival is held on the fifteenth night of the seventh month in the Chinses calendar, at the end of August or beginning of September. It is a festival to honor the dead who were not given a proper funeral so they can be released from the underworld on this day. Taoists burn incense and joss paper, leave food offerings, and provide entertainment and music to make them happy.

Mid-autumn Festival: This festival is held on the fifteenth day of the eighth month of the lunar calendar, mid-September to the beginning of October, around the autumn equinox. It is a traditional ethnic celebration of Chinese and Vietnamese.

Double Ninth-Day Festival: This is a day of reverence for ancestors. This festival is held on the ninth day of the ninth month in the lunar calendar; in October. This day is celebrated by climbing a mountain and drinking chrysanthemum wine to avoid evil spirits and misfortune.

Summary

Taoism is an aboriginal religion-philosophy attributed to Lao Tzu and began with the writing of the Tao-Te-Ching, or the Book of the Way, in 300 BC. Taoism is a system of folk religion and philosophy with no real founder, as well as no date of founding. The teaching of the Tao-Te-Ching came to fruition by thinkers like philosopher Yang Zhu, who was born in 440 BC and died in 360 BC.

Taoism originally started as a philosophy; a couple of hundred years later, Zhang Daoling included religion for more participation and expansion of Taoism. The reason he added religion was for competition against Buddhism and Confucianism. Taoism is not interested in death; it's part of the cosmic cycle. This means being born out of emptiness and going back to a vacuum, and life and death are two aspects of reality, the unchanging Tao.

Taoism developed in ancient China, so one cannot ignore respect to ancestral belief, which is part of the culture and history of Chinese, whose ancestor worship goes back to 3000 BC.

In Taoism, there is no God and no fear component. The civil government keeps society under control based on the law. In case the empire is not working for the well-being of the nation, the nation stands up to remove the emperor and bring in a new emperor.

CHAPTER 11

Shintoism

Izumo Taisha is one of the most ancient Shinto shrines. It stays close to the beach which is named Taisha Beach in Izumo City, Shimane, Japan. The gods from around Japan descend here once per year (Copyright iStock by Getty Images/ credit: Alexander Pyatenko and Coolvectormaker).

Shintoism is a purely Japanese polytheistic religion and revolves around the *kami* (English: spirits or gods). The aboriginal religion, or old Shinto, owes nothing to outside sources; the entire religion is an independent development of Japanese thoughts. Shintoism has no founder, no revered teacher, and no primary sacred text, like the Koran or the Bible.

Shintoism has two highly revered scriptures: Kojiki and Nihon Shoki (Nihongi). Kojiki, or Record of Ancient Matters, is a repertory of the old myths, legends, and the ancient history of Japan. The Nihon Shoki, or Chronicle of Japan, is a work of similar scope based on written literature. It describes the myth of the divine origins of Japan.

Emperor Temmu, who reigned from 672 to 686 CE, was concerned that the ancient legends of celestial origins of the imperial line and Japanese people were corrupt and would be lost, so he ordered them memorized. Later, Emperor Gemmyo (or Genmei), who reigned from 707 to 714 CE, ordered the memorized version written. The first written versions of Kojiki and Nihon Shoki go back to 712 CE and 720 CE.

The doctrine of Shintoism is that Japan is the land of the gods. Its inhabitants are the descendants of gods. Japanese indigenous religion came into practice before the Japanese language, the same as other indigenous religions around the world, and had no name until they interfaced with Confucianism, Taoism, and Buddhism.

Shinto is derived from two Chinese words: Shen-tao, Shen (Chinese: deity), and tao (Chinese: way or path). The Japanese version of Shinto comes from "the way of gods" or "kami no michi," which means a way of kami. (Kami means a god, a superior being.)

Shintoism follows the ancient folk belief that the natural world has spiritual powers, and they think that the spirit called kami lives in people, stones, plants, animals, rivers, mountains, the dead, and rocks. Shintoism is very much like animism, and both religions believe that spirits exist in nature. Shintoism is known as Japanese animism; it is part of the primitive religion of hunter-gatherers.

According to Shinto myths and folklore, the Japanese islands were created by the central deities (or kami); Izanami (Izanami no

Mikoto, meaning She Who Invites) is a goddess of creation and death, and Izanagi (Izanagi no Mikoto; meaning He Who Invites). They allegedly dipped a spear into the primordial sea, and the Japanese islands were created. They also created more than eight hundred gods, the deities of Shinto, and, most importantly, created the kami Amaterasu, the sun goddess.

Amaterasu's grandson Ninigi became the first king, and he was the great-grandfather of the first emperor, Jimmu. Amaterasu's shrine is located at Ise (formerly called Ujiyamada), and is the most important shrine in Japan. This myth established a divine link between all the future emperors and the goddess of the sun. The most prominent gods (kami) are in the following order:

1. Izanagi and Izanami, or He Who Invites and She Who Invites, are the central deities who created the island of Japan, the Japanese, and the world. They are also the first man and woman.
2. Amaterasu Omikami, the great divinity illuminating heaven, is the sun goddess, the most important deity of the Shinto religion.
3. Kagutsuchi is the god of fire, also known as Homusubi. He was the son of Izanami, and the father of eight warrior kami.
4. Hachiman is the god of war; he sent the kamikaze or divine wind to protect Japan from the invading fleet of Mongol ruler Kublai Khan, for which he earned the title Protector of Japan.
5. Fujin is the god of wind. The island of Japan has historically been devastated with storm and typhoons, which caused a lot of damage to communities. Therefore, Fujin is respected and feared for its power.
6. Inari Okami is the god of rice, the protector of agriculture and bringer of prosperity. Forty thousand shrines large and small across Japan are dedicated to Inari.
7. Shichifukujin is the Seven Lucky Gods, also known as the Seven Gods of Good Fortune and the Seven Gods of

Happiness. These deities have a diverse origin; some from India, China, and Japan. Ebisu is Japan's only indigenous Shinto tradition, and the Seven Lucky Gods are the following:

    a. Ebisu, the god of luck
    b. Daikoku, the god of wealth and prosperity
    c. Benten (or Benzaiten), the goddess of love and beauty
    d. Bishamon, the god of pleasure and war
    e. Fukurokuju, the god of happiness and longevity
    f. Jurojin, the god of longevity
    g. Hotei, the god of generosity

8. Kotoamatsukami is the heavenly god, a collective name for the first gods who came at the time of creation.

The earliest civilization in Japan was during the Yayoi period, between 400 BC and 250 CE. Around 400 BC or previous, migrants began to reach Japan through the Korean Peninsula. The immigrants integrated with the indigenous people of Japan. The new arrivals came with different cultures, religions, and industries, like weaponry, new armor, and bronze and iron. Sometime during the Yayoi period, the Japanese authorities gave the name of Shinto to their old state religion to distinguish it from other religions.

Leaders

The first king of Japan, King Ninigi, is the grandson of the supreme Shinto deity Amaterasu, the sun goddess. King Ninigi is the great-grandfather of the first emperor, Jimmu. This established a divine link between future emperors as the most prominent leaders of Shintoism and the gods. In addition to emperors, there are a few other essential leaders and influential scholars, as shown below.

Jimmu: Jimmu or Jimmu Tenno (original name Kow-Yamato-Iware-Hiko No Mikoto) is the first legendary Japanese emperor and founder of the imperial dynasty. In 607 BC, he led an expedition along Japan's island and subdued all the tribes and established his center of power at Yamato. In 1890 CE, a Shinto shrine was erected in Jimmu's memory at the site where he was believed to be buried at Unebi.

Tenmu: Despite the earlier appointment of Crown Prince Oama Tenji's brother by the court as successor in 664 CE, Emperor Tenji named his favorite son, Prince Otomo, as the successor in 671 CE. After Emperor Tenji's death, a civil war broke out to determine succession to the throne. Crown Prince Oama won the civil war and renamed himself Tenmu, the fortieth emperor, and he was assigned the title of Tenno, which means emperor of Japan. Emperor Tenmu was concerned that the ancient myths of celestial origins of the imperial line and Japanese people, Kojiki and Nihon Shoki, would be corrupt and lost, so he ordered them memorized. Later, his descendants put them in writing. He moved the capital from Otsu to Yamato Province and renamed it Asuka. Emperor Tenmu converted to Buddhism in 685 CE and paid stipends to Buddhist priests and nuns.

O no Yasumaro: O no Yasumaro, the son of O no Honji, was an active fighter in the Jinshin War, where Tenmu became the emperor. O no Yasumaro authored two scriptures that were originally commissioned by Emperor Tenmu to be memorized: The Kojiki, or An Account of Ancient Matters, in 712, and the Nihon Shoki, or the Chronicles of Japan, in 720 CE. He died in 723 CE, and was presumed welcomed to the World of Darkness by the kami, the spirit of Japan.

Motoori Norinaga: Motoori Norinaga was born in 1730 CE in Matsuzaka, Japan, and died there in 1801 CE. His home has been preserved as a museum and a Japanese national monument. He lost his father at the age of eleven, and his mother encouraged him to study medicine. He also studied Chinese and Japanese philology. At some point, the National Learning movement (Kokugaku) influenced him; this emphasized the importance of Japan's literature and classic Shinto. Motoori Norinaga rejected the Confucian and Buddhist beliefs and created a foundation for the modern Shinto revival. He wrote extensive commentaries and annotations on the

Kojiki. His advocacy of Japanese history and religious learning led to the Meiji Restoration of Shinto to imperial Japan.

Hirata Atsutane: Hirata Atsutane was born in 1776 CE, in Akita, Japan, and died there in 1843 CE. Due to financial difficulties, he moved to Edo, modern-day Tokyo. He was trained in the reading of Chinese texts and neo-Confucianism and later turned to Shintoism. He followed the Motoori Norinaga way of thinking and a year after Motoori Norinaga's death, his discipleship was officially acknowledged. He became a leader of the National Learning movement (Kokugaku). He was critical of the Tokugawa feudal regime, in which the emperor is a powerless symbol. His works include Koshi, or Ancient History; Koshicho, or Reference to Ancient History; Koshiden, or Commentaries on Ancient History; Tama no Mahashira, or the Real Pillar of the Spirit; Shinkishinron, or A New Discourse on Kami; and Honkyogaihen, or Additional Teachings of the Central Tradition. After his death, his disciples contributed to the establishment of the new government to the Meiji Restoration.

Emperor Meiji: Meiji Tenno (original name Mutsuhito) was born in 1852 CE in Kyoto, Japan, and died in 1912 CE in Tokyo, Japan. Emperor Meiji reigned Japan from 1867 CE to 1912, and consolidated Japan's government under imperial rule. During his reign, he oversaw a period of modernization, industrialization, and reform called the Meiji Restoration. He took an active role in the execution of the Sino-Japanese War and the Russo-Japanese War. In 1910 CE, he proclaimed the annexation of Korea to Japan.

Emperor Hirohito: Hirohito's original name is Michinomiya Hirohito. He was born in 1901 CE in Tokyo, Japan, and died in 1989 CE in Tokyo. His posthumous title is Showa, which means Bright Peace or Enlightened Harmony. During World War II, he sided with Germany and invaded part of China and followed the policy of expansionism. When Japan was facing defeat in 1945 CE, he announced Japan's acceptance of the Allies' terms of surrender.

Consequently, Japan became a constitutional monarchy and had to relinquish the imperial claim to divinity. This constitution directly challenged the Shinto story, which tells that the emperors of Japan are descended continuously from the first emperor, Jimmu,

Amaterasu Omikami's great-grandson. Based on this constitution, Shinto is no longer a state religion.

Priests: Priests in Japanese are called Kannushi (originally pronounced Kamunushi or Shinshoku), meaning the gods' employee. Shinto has close to 85,000 priests who live on the shrine grounds. Shrines are homes to kami, and every shrine has an inner hall where kami is present, which is only entered by priests. Both men and women can become a priest, and they are allowed to marry and have children. Priests maintain the shrines and perform Shinto rituals. Usually a young unmarried woman (could be the daughter of a priest) called a miko assists the priest in ceremonies. She wears a red skirt and white jacket, and her function is to perform a sacred dance called Kagura to honor kami. In the old days of Japan, the holy and the secular imperfectly differentiated from each other.

The Mikado, a title for the Japanese emperor in ancient Japan, was a high priest as well as the emperor of Japan. The chief officials of Shinto were appointed from the hereditary clan called the Nakatomi clan. Also, from the same clan were selected the principal minister of state and the imperial consort. The Imbe and the Urabe clans have high priestly functions too. During the Meiji period, Shinto priests became state officials.

Scriptures

Shinto has no primary sacred text like the Koran or the Bible but has two historical scriptures, Kojiki (712 CE) and Nihon Shoki (720 CE). Compilation of these scriptures had a twofold benefit. First, to protect the ancient myths of celestial origins of the imperial line and Japanese people, and the second, political intention was to establish the superiority of Japan and Japanese people over the world.

Kojiki: The Kojiki (Japanese: Record of Ancient Matters) scripture was compiled from oral tradition. The contents of the Kojiki are myths, legends, ceremonies, divinations, customs, magical practices, and historical accounts of imperial court from the creation of Japan to the reign of the Empress Suiko in 628 CE. Since Japanese writing was not developed, the original scripture was written using

Chinese characters to represent Japanese sounds. Motoori Norinaga (1730–1801 CE) rewrote the Kojiki with complete commentaries in forty-nine volumes.

Nihon Shoki: The Nihon Shoki or Nihongi (Japanese: Chronicles of Japan) scripture was compiled from oral tradition. The content of the Nihon Shoki is myths and legends concerning the Shinto gods, and is the oldest official history of Japan from the creation of Japan to the reign of the Empress Jito in 697 CE.

Denominations

Shintoism is a purely Japanese religion and did not spread outside Japan. Japanese indigenous religion came in practice before the Japanese language, the same as other indigenous religions around the world, and had no name until they interfaced with Confucianism, Taoism, and Buddhism. Emperors were considered descendants of the sun goddess Amaterasu. Shinto has a few categorical denominations.

Jinja: Jinja Shinto is the oldest Shinto denomination and is thought to date back to prehistoric times. This denomination is considered to be the core of Shintoism. Shinto Jinja is associated with the eighty thousand shrines that give their loyalty to the Association of Shinto Shrines (Jinja Honcho).

Kyoha: Kyoha, or the Sect Shinto denomination, started in the nineteenth century. This denomination has thirteen major independent sects that are officially recognized by the Japanese government.

Minzoku: The ordinary people at local shrines practice Minzoku, or the Folk Shinto denomination. This sect is a combination of several religions, like Buddhism, Taoism, and Confucianism, and any other national reforms that do not institutionalize it. These people worship the local kami at local shrines called kamidana, or god shelf, and it includes a fountain with a stone altar. In this altar, there are two bowls; one to drink from and cleanse the mouth, and the other to wash hands in.

Koshitsu: Koshitsu Shinto, or Shinto of the Imperial House, is a state-sponsored Shinto. Emperor Meiji founded this denomination in 1800. This denomination values nationalism, patriotism, expansionism, and obedience to the emperor and government. Koshitsu Shinto was abolished after World War II.

Beliefs

Shinto means the way of the gods, and their belief centers on spirits called kami, or a divine being, live in a natural world, even dead, and especially in and throughout Japan.

Deities: The gods of Shinto are known as kami. Kami are very close to human beings and respond to their prayers. Everything in nature—birds, rocks, social beings, animals, trees, plants, mountains, moon, sun, wind, and streams—is kami. Shinto is a polytheistic and animistic religion. Tradition says that there are around eight million kami; some kami are very important, and some are not. The most critical kami has many myths. The sun goddess is the greatest of all kami. The sun goddess is the kami of the Ise Shrine and the ancestor of the imperial family. Japanese are not obligated to worship any specific kami.

Creation: Shinto believed that the universe always existed as a vast, oily, reedy ocean and sky. The creator gods are Izanagi (Japanese: He Who Invites) and Izanami (Japanese: She Who Invites). Izanagi placed his spear into the ocean, and the tip of his spear felt mud. This mud created the land of Japan, and on this land, the divine couple descended. This godly couple gave birth to all the other deities, the Japanese people, and the rest of the world. Izanami died, and Izanagi, after seeing his dead wife, bathed in the sea to purify himself. While he was washing, the three most important deities came into being. The sun goddess, Amaterasu, came out of his left eye. The moon god, Tsukuyomi, came out of his right eye. And the storm god, Susanoo, came out of his nose.

Ethics: Shinto has very little in the form of a code of ethics. Whatever is in Shinto's ethical system was inspired by Confucian values, and from the idea that both nature and humans are good, and humans are born in a state of purity. Shintoists try to live by way of the kami in such a manner as to keep a good bond with the kami. Based on Shinto texts, the kami are not perfect; they make mistakes and do wrong things. The bottom line is Shinto ethics are to stimulate purity and harmony in all spheres of life. However, when evil enters and spreads in human beings, in a way like a disease, and reduces their abilities to resist temptation and causes doing a wrong act or sin, the person must act for purification. In Shinto, there are two times a year grand purification ceremonies known as Ohoharahi. They make an offering of sacrifice and throw it into the rivers and sea to carry with them the sins of the people.

Scriptures: Shinto has no primary sacred text, like the Koran or the Bible. Shinto, up until 400 BC, did not have any writing at all. However, Shinto has two historical highly revered scriptures, compiled in the eighth century in response to Chinese religions and Buddhism. These scriptures are Kojiki and Nihon Shoki (Nihongi). Kojiki, or Record of Ancient Matters, is a repertory of the old myths, legends, and the ancient history of Japan. The Nihon Shoki (Chronicle of Japan) is a work of similar scope based on written literature, which describes the myth of the divine origins of Japan.

Death and the afterlife: The old traditions viewed death as the source of impurity, like Izanagi in the creation myth. The priest will not be present during the funeral ceremony or allow the corpse in the shrine. Also, the family closes and covers their home altars to keep the spirit of the dead out. Immediate family members mourn as a natural response to death. They show their intense grief on a specified day, but mostly they express their grief in a way that holds the deceased in the highest respect. A process of purification requires the spirit to be able to move into the spirit world. The spirits live in another world called the other world of heaven, which is the most sacred world. Those who cannot become ancestral kami on earth spend forever in Yomi, the world of darkness. In Shinto, the other world is not seen as a reward or punishment, simply as a place where the spirits reside.

Rituals and Customs

Shinto does not have a creed or divine scripture. Shinto depends on ritual to convey religious thoughts and feelings. Shintoists believe that kami will appreciate and respond to their rituals. The ritual can take place at home or a shrine, and the rituals are always the worship of kami.

Purification: Purification (Japanese: Harai or Harae) is a ritual that removes sin, pollution, and impurity (Kegare) from a person or object. Purification can be performed in many ways, including cleansing by water or saltwater, or prayer from a priest. When entering a shrine, there are specific ablutions that must be followed.

1. *Harai-Gushi and Ohnusa* is a method to remove Harai from a person to an object utilizing an object. A priest will wave a wand attached to the end of the rod, a strip of paper over a person, and the polluted Harai-gushi will be destroyed later.
2. *Imi* means mourning period and is a preventing method. For example, the members of the family of someone who has passed away will not be allowed to enter the shrine.
3. *Misogi Harai* is the ancient method of purification: submerge oneself under the water, the same as Izanagi did.
4. *Oharae*, or the ceremony of great purification, is a method the Japanese perform every year in June and December. With this purification, all of Japan is purified from sin and pollution.

Shrines: Shinto has close to eighty thousand small and large public shrines. Ancient Shinto had no shrine buildings. In response to the arrival of Buddhism, Shinto developed permanent structures dedicated to kami. Shinto shrines are marked by a gateway (Japanese: torii) that stands at the entrance of the shrine. Shrines have an inner hall, where a sacred symbol called shintai or kami body is present, which is only entered by a chief priest. Anyone is welcome to visit public shrines; a visitor will come first to an ablution basin, where he or she washes their hands and rinses their mouth and makes a small offering. During the ancient Shinto, clans took care of the shrines.

Worship: Shinto prayer or devotion (norito) has no set schedule; Shinto worship as they feel the need, in hopes for good fortune. Worship in Shinto is more critical than belief, and the Japanese learn from birth the habit of reverence to kami. Shinto's glorifications or worships are highly ritualized and follow strict order and control, and can take place at home or in the shrines. The chief kannushi, or priest, recites prayers following ancient Shinto prayers. The prayers were compiled in the early tenth century, according to the old belief that spoken words had spiritual power.

Shinto rituals should be carried out in a spirit of honesty, purity, and truthfulness. Shinto worshippers, before entering the shrine, purify and put on new clothes. From the torii or ritual gate to the shrine is a path and along the path, there is always running water of some form. The worshippers wash their hands and take a little water in their cupped hands and gargle. They never touch the source of water with their lips. The worshipper drops a small coin in a coin box and strikes a gong or rings a bell to gain kami's attention. They leave a small offering, such as coins or a little sake, for kami. Worshippers buy prayer or fortune slips called omikuji, maybe blessings or curses. They leave the evils behind and take the fortunes to their home altar or kamidana (god shelf). Home altars are built above eye level and include a symbolic object like stone, mirror, or jewel to act as a physical home for kami. Home altars are decorated with a strip of paper or cloth called ofuda, featuring the name of a kami issued by a shrine.

Sacred music and dance: Kagura means gods' entertainment (also called Kamiasobi in ancient times). It is a type of dance to appease and energize kami rather than to praise them. Kagura includes music played on an ancient Japanese stringed instrument called a yamatogoto, and dragon flutes. Kagura goes back to Japan's origin myth, when kami used music and dance to coax Amaterasu out of the cave and restore light to the world at dawn. Kagura divides into two forms, Mikagura and Okagura. The Mikagura is performed in the imperial court and originated in kinkashinen (religious feasts). The Okagura is performed at shrines during folk ceremonies.

Marriage rituals: The Shinto wedding is a reasonably small affair; the ceremonies traditionally occur in the presence of family, close friends, and a priest at a shrine. The bride wears a white kimono, and the white symbolizes purity. The ceremony consists of ritual purification, exchanging vows, and prayers offered to the couple to have good luck and protection of the kami. The couple drinks three sips of sake offered by the miko, the shrine maiden. They exchange rings, the groom reads words of commitment, and makes an offering to the kami. This is followed by a sacred dance performed by the miko and ends by ritual sharing of sake by everyone present. These ceremonies are relatively recent, after the wedding of Crown Prince Yoshihito and Lady Sadako Kujo[100] in 1900. Before that, the celebrations were different and long.

Birth and infancy rituals: Shinto, like all religions and cultures of the world, marks times in human life: death, marriage, and birth. The Shinto[101] emphasis is on the importance of the birth ritual more than death or marriage, and includes the following rituals.

1. *Obiiwai (the first dog day)*: In the fifth month of pregnancy, a cotton belt will be tied to the pregnant woman to protect the baby. The family goes to the shrine to pray for ease of birth. It's called the dog day because dogs have natural deliveries.
2. *Oshichiya (the seventh night after birth)*: The father announces the name of the baby to the family with a ritual. The entire family will have a celebration dinner.
3. *Omiyamairi (the first visit to a shrine)*: This is a prominent ritual where the family takes the baby to a local shrine to show their new baby to kami. This ritual takes place when the boy is 31 days old and the girl is 33 days old. The baby is dressed up in a detailed kimono gifted from the mother's family and a grandmother holds the baby. The priest offers a prayer for the good health and happiness of the baby; family and friends bring gifts.
4. *Okuizome (the first bite of food)*: About 120 days after birth, a traditional, ancient full-course meal is served to the baby. The parents take turns and feed the baby; it's considered to be an offering of power.
5. *Hatsuzekku for girls (the first hinamatsuri, meaning girl's day)*: The first March 3 of her life is the first hinamatsuri, and it's very special to the family. They serve sweet sake to share the happiness of having a baby girl. This celebration goes back to the eighth century CE.

6. *Hatsuzekku for boys (the first boy day)*: The first May 5 of the boy's life is an extraordinary day for the family. The family decorates inside the house with samurai dolls and outside with koinobori (carp streamers) to show that there is a boy in the family. This ritual goes back to the seventh century CE.
7. *The first birthday*: Traditionally, the Japanese make the walk alone carrying four pounds of mochi (a sticky rice cake that is a very sacred food in Shintoism) on the baby's back. Doing this, the family wishes to give holy power to the baby's life.
8. *Shichigosan (7-5-3 festival) for girls*: The celebration of Shichigosan takes place on November 15 or the nearest Sunday. They all dress in kimono to visit a shrine with the family and have a prayer at the age of three, five, and seven years to give thanks to the gods for a healthy life now and pray for success and health in the future. This tradition goes back to the eighth century CE.
9. *Shichigosan (7-5-3 festival) for boys*: The celebration of Shichigosan takes place on November 15 or the nearest Sunday. They all dress in kimono, like noble samurai, to visit a shrine with the family and have a prayer at the age of three, five, and seven years to give thanks to the gods for a healthy life now and pray for success and health in the future. The boys hold a small sword imitating a real katana. This tradition goes back to the eighth century CE.
10. *Seijin Shiki (Adulthood Day)*: Seijin Shiki is one of the most important celebrations for all those who reach the age of twenty years old. Japanese who have had their birthday before January 15 attend a shrine to give thanks to kami. They all dress up in the kimono and join the ceremony in their hometown, and they become a real member of society.

Festivals and Holidays

Shinto originated in a farming society of Japan, and with the influence of Chinese religions, most of their festivals (or matsuri) are tied to the agrarian calendar. *Matsuri* means welcoming the descending gods or inviting down the gods. Shinto festivals are physical events and include processions, dancing, feasting, sumo wrestling, and dramatic performances and are colorful, loud, and aromatic with smells of food. There are many festivals based on regions and shrines, and the following are the most important.

Oshogatsu, the New Year festival, is observed from December 29 through January 3. At this festival, the Japanese visit a shrine to express appreciation for kami in the past year and the year to come. The first shrine they visit is called Hatsumode. In 1948, January 1 was established as a national holiday.

Hana Matsuri (Buddha's Birthday Festival) is held every year in Japan in April. It celebrates in a ceremony called Kanbutsu-e (Flower Festival) to commemorate the historical Buddha's birthday.

Oharae, or the ceremony of great purification, is a method the Japanese perform every year on June 30 and December 31. With this purification, all of Japan purifies from sin and pollution.

Summary

Shintoism is a purely Japanese polytheistic religion and revolves around the kami (English: spirits or gods). It has no founder, no revered teacher, and no primary sacred text. Shintoism is an independent development of Japanese thoughts. It has two highly revered scriptures: Kojiki and Nihon Shoki (Nihongi). Kojiki, or Record of Ancient Matters, is a repertory of the old myths, legends, and the ancient history of Japan. Shintoism is known as Japanese animism, and is part of the primitive religion of hunter-gatherers. Sacred music and dance are another element of primitive religion in Shintoism called Kamiasobi.

Kami (gods) are very close to humans, and respond to human prayers. Some Shintoists believe there are about eight million kami in Japan. Shintoists think that kami will appreciate and respond to their rituals. The ritual can take place at home or a shrine, and rituals are always the worship of kami. Shintoists believe that a purification ritual cleanses sin, pollution, and impurity from a person or object.

Purification can be done in different ways: by water, saltwater, or prayer from a priest.

Shintoists do not talk much about death and the afterlife. Shintoists view death as the source of impurity, like Izanagi in the creation myth. The priest will not be present during the funeral ceremony or allow the corpse in the shrine. They perform a purification for the corpse so that the spirit is able to move into the other world. In Shinto, the other world is not seen as a reward or punishment; it is a place where the spirits reside. Therefore, there is no element of fear in Shintoism.

CHAPTER 12

Christianity

Vintage engraving from 1876 from an original showing unloading of the crusaders in Damietta Port, Egypt (Copyright iStock by Getty Images/credit: Duncan1890 and Sylverarts).

Christianity sprang from Judaism and is one of the Abrahamic religions. The beginning of Christianity starts with the magical birth of Jesus around 33 BC, and includes his adult life, teaching, death, and resurrection (Arabic: an-Nasira, Hebrew: Nazerat) in a historic city of lower Galilee, in northern Israel, during the height of the Roman Empire. Those Jews and non-Jews, who called themselves the followers of Jesus, designated a new religious community that was first called Christ (Greek: Christos) in the Syrian town of Antioch, possibly 35 to 40 years after the death of Jesus. Christ is a title, not a proper name. It comes from the Greek word *Christos*, meaning the Anointed One of God, a Greek translation of the Hebrew title Messiah.

The Christian faith has become one of the largest religions in the world; according to a 2015 statistic (Pew Research Center), there are 2.3 billion Christians around the world.

At the beginning of the Roman Empire, the largest empire in the world expanded Christianity in Europe and part of Asia, and later from around the fifteenth century to the present time, European colonialism became a significant force to spread Christianity around the world.

Its largest denominations are the Roman Catholics, the Eastern Orthodoxies, the Protestants, and the Oriental Orthodoxies, and there are many more sub-denominations.

Historically, religions rapidly expanded when adapted and supported by a king or an emperor. Regardless of whether Christianity was invented by the Roman Empire or not, it became the official religion of the Roman Empire in 323 CE. The Roman kingdom started around 753 BC, followed by the Roman Republic around 509 BC, and finally, the first empire of Rome was announced by Emperor Augustus in 27 BC.

Religion was a vital political tool for Rome's internal and external governance, development, expansion, and diplomacy from kingdom to republic to empire. The Roman Empire, with a new religion, was able to rapidly spread the Christian faith to all of Europe and parts of Asia and Africa. The dominance of the pope and the Catholic Church was the only stable force in Europe. The pope had the power to remove a king or emperor and install a new king or emperor of his choice.

The Catholic Church and Christians were also at the forefront of the Crusades and many other atrocities around the world.

In the meantime, Europe was on the ladder of development in philosophy, art, science, music, literature, architecture, and medicine. The Catholic Church was in control of events until the age of Enlightenment, from 1637 to 1789, when their power reduced and gradually. Separation of church and state was implemented in Europe and America.

Leaders

Christianity has a long list of prominent leaders. This book will start with Jesus Christ and some of his apostles.

Jesus Christ: Jesus's date of birth and time of the crucifixion are not stated in the gospel, even though the largest empire in the world crucified him. Most scholars agree on his year of birth as 32 BC[102] in Bethlehem, according to Matthew and Luke, and he was crucified in 1 AD in the old city of Jerusalem. Jesus is also called Jesus of Nazareth, Jesus Christ, or Jesus of Galilee. His mother's name was Mary, a virgin when Jesus was conceived. She was found to be with child of the Holy Spirit. There is no information about her birth, death, and no burial place is acknowledged. His father's name was Joseph. However, according to Matthew and Luke, Joseph was Jesus's *legal* father. Mary pledged to be married to Joseph when Joseph discovered Mary was pregnant. He was a righteous man, and he did not want to expose her to the people's disgrace and death by stoning; he wanted to divorce her quietly. The angel of God came to him in a dream and told him, "Joseph, son of David, do not be ashamed to take Mary as your wife, because what is conceived in her is from the Holy Spirit. She will give birth to a child, and you have given him the name Jesus" [Hebrew: Joshua, Yeshua, or Isa] because he will save his people from their sins." When Joseph woke up in the morning, he did what the angel told him (Matt. 1:18–25).

Joseph and Mary were peasants, and the peasant life was quiet and straightforward. Therefore, there is not much information about his childhood life or his early adulthood. Jesus was never married, and there is no information about his marital status. Luke 2:41–52 explains that Jesus as a child was talented and mature, but Luke did not say if Jesus was literate or not.

At the time of Jesus, the majority of people were illiterate. Jesus was poor, and from a peasant family, he was a trained carpenter (Mark 6:30, Matt. 13:55). There isn't a single writing or scripture from him, nor other documentation about his literacy. We can conclude that Jesus was illiterate.

Baptism is a traditional ritual and purification rite of Judaism. Therefore Jesus, as a Jew, was baptized (Matt. 3:13–17) by Saint John the Baptist (born first decade BC, in Judaea, Jerusalem, died 30–36 CE). Shortly after baptism, Jesus became a preacher, a healer, and performed miracles. Jesus's first miracle was that he changed water to red wine at a wedding in Cana (John 2:1–11). Jesus healed a government official's son in Capernaum while the boy was close to dying (John 4:46–54), and he treated a person with paralysis at Bethesda (John 5:5–15).

Jesus didn't have a long history of preaching, healing, and miracles. It lasted less than one year after his baptism. During this time, he attracted a few followers, and he went to observe Passover in Jerusalem. At the time he was in Jerusalem he was arrested, tried, and executed. His followers hoped for the launching of a holy war against the Roman Empire's occupation of Judea (Hebrew: Yehudah). Still, he taught them nonresistance and to show love to enemies.

Three days after his execution, he rose from death; his disciples were convinced that he was still alive and had appeared to them. The first appearance of Jesus was to Mary Magdalene (John 20:14–16, Luke 24:39) and then the other disciples. Most importantly, this confirmed Jesus's testimony about his resurrection (Rev. 1:18).

Jesus was executed by order of Pontius Pilate. Pontius Pilate (full name in Latin: Marcus Pontius Pilatus) was born in Rome in 12 BC and died in 38 CE. He was the Roman prefect, or governor, of Judaea from 26 to 36 CE under Emperor Tiberius. He was one of the cruel, oppressive, and powerful governors of the Roman Empire who had executed men without proper trial. History is a witness of evil emperors, kings, and governors; when they killed criminals, traitors, and enemies, the bodies were hung on public display and became a food source for birds and crows. The dead bodies were watched by professional army personnel so that nobody could go close to them until they turned into clean bones. Under these circumstances, Jesus did not resurrect. Besides, his bones were dumped. Therefore no burial site or tomb was created for Jesus.

The gospel does not claim that Jesus was not a Jew. Jesus was born from a Jewish mother, Mary, and he was born a Jew. Jesus, as a Jew with all his Jewish disciples, went to a synagogue to observe Passover in Jerusalem. There he was arrested and executed as a Jew. After his execution, the disciples believed in him, which ultimately led to a new religion. As described earlier, Jesus was preaching, healing, and performing miracles for less than a year. He was not able to create a large community of followers except a few disciples, and therefore, Jesus was not considered the founder of Christianity. One of his apostles, Saint Paul, (whose original name was Saul of Tarsus of Cilici, now modern-day Turkey) is considered the founder of Christianity.

Paul of Tarsus: Saint Paul the Apostle, whose original name was Saul of Tarsus in Cilici, now modern Turkey, was born around 4 BC, and died in Rome around 64 CE. There is not much information elsewhere other than the Bible. He was converted to Christianity on the road to Damascus in an extraordinary way.

> As he was approaching Damascus, a light from heaven suddenly shone around him. He dropped to the ground, fall asleep, and heard a voice saying to him, "Saul! Saul! Why are you persecuting me?" "Who are you, Lord?" And the voice replied, "I am Jesus, the one you are persecuting! Now get up and run into the city, and you will be told what you must do." (Acts 9:3–6)

This story suggests that Paul was involved in Christian life from the beginning of Christianity. He was a Roman citizen and a Roman prosecutor who was prosecuting lawless people. After his conversion to Christianity, Saul of Tarsus became a passionate missionary. He traveled throughout the eastern Roman Empire, Asia Minor, and present-day Greece as a Roman citizen with Roman authority, spreading the good news that Jesus would come back soon to rule and reign on earth.

In the New Testament, there are fourteen letters attributed to Paul. The scholarly consensus at present says seven out of the fourteen letters were written by Paul.[103] Paul is believed to be the founder of Christianity and one of the most exceptional leaders of Christianity. It is thought that he died or was executed in Rome around 64 CE under the order of Emperor Nero Claudius, who reigned Rome from 37 to 68 CE. There is no information on where he is buried.

Saint Augustine: Saint Augustine, also called Saint Augustine of Hippo, was born on November 13, 354, in Souk Ahras, Algeria, and died on August 28, 430, in Annaba, Algeria. He was born to a Christian parent; his mother was a devout Christian. Based on his word, he was sinful, self-centered, selfish, and hedonist. Later, in his middle age, he dabbled in religious groups; he fell in with the Manichaeism, a Persian religious movement started by Manes (or Mani, born in Gundeshapur, Iran, in 216 and died in 276 CE).

Soon, he realized he was not going to get what he wanted, so he moved to Milan and met with Saint Ambrose, the bishop of Milan, to be baptized by him. In 391 CE, he became a co-bishop and later bishop of Hippo. His work was *Confessions* and *The City of God*. *Confessions* is his autobiography. He devised the doctrine of original sin. He said everybody is born sinful, which means the human is born with the built-in impulse of doing bad things. This doctrine of Saint Augustine of Hippo became part of Roman Catholic doctrine by the Councils of Trent in the sixteenth century.

Constantine the Great: Constantine the Great is also called Constantine I (Latin: Flavius Valerius Constantinus). He was born in 280 CE in Nish, Serbia, and died in 337 CE in Izmit, Turkey.

Historically creating a religion and faith or using the existing religion and belief is the best tool for success and victory. He learned from his predecessors that converted to Christianity to use this to his political advantage. He fought many battles, and especially the Battle of Milvian Bridge outside Rome in 312 CE, under the banner of Christianity, which was a crucial moment of his success; Constantine became sole ruler of the Roman Empire.

In 313 CE, he provided the Edict of Milan, which permitted Christianity legal status. After Constantine defeated the eastern empire, Licinius renamed Byzantium Constantinople in 324. He created laws for the Church and its officials' fiscal and legal privileges and protection from the civic burden. He donated the imperial property of the Lateran to the bishop of Rome where a new cathedral, the Basilica Constantiniana, soon arose. It was another way to get complete support of the Church. Constantine called Church officials to the council of Nicaea in 325 CE to eliminate the conflict within the Church. He was the first speaker of the council. Out of this council came the Nicene Creed, which affirmed that Jesus was a divine being. Constantine created the most extensive Christian empire, which included Europe, part of Africa, and part of Asia. Finally, after many years of using Christianity for his political game, he converted to Christianity on his deathbed in 337 CE.

Saint Thomas Aquinas: Saint Thomas Aquinas is also called Angelic Doctor and was born about 1225 at the castle of his father in Roccasecca near Aquino, Terra di Lavoro, the kingdom of Sicily, Italy. Aquinas died in 1274 in Fossanova, near Terracina. He is recognized as one of the most influential philosophers and theologians by the Roman Catholic Church. The modern Roman Catholic theologians do not find Thomas Aquinas congenial. He worked hard to reconcile faith and reason. His most significant works are the Summa of Theology, Summa contra Gentiles, and commentaries on the Bible and Aristotle.

John Wycliffe: John Wycliffe was born in 1330 CE in Yorkshire, England, and died in 1384 CE in Lutterworth, England. After completion of his education at the University of Oxford, John Wycliffe became an English philosopher, theologian, anti-authority of Church and pope, and believed the translation of the Bible into English is best for the people to understand Christianity. He oversaw the first widespread English translation of the Bible, known as the Wycliffe Bible. Wycliffe was a brilliant and heroic scholar and cleric who believed in simple term separation of state and church. One of his revolutionary works was *On Civil Dominion*, in which he wrote, "England belongs to no pope. The pope is but a man, subject to sin, but Christ is the Lord of lords, and this kingdom is held directly and solely of Christ alone."

In 1401, a new law was passed and ordered that the heretics be burned and it was illegal to read the English Bible or any Wycliffe works. Forty-four years after John Wycliffe's death, his bones, including all his writing, were burned in a field of execution by order of Pope Martin V, and the ashes scattered in the river Swift near Lutterworth.

Martin Luther: Martin Luther was born in 1483 CE in Eisleben, Germany, and died in 1546 CE in Eisleben, Germany. Luther graduated from the University of Erfurt, one of the top universities in Germany. He took courses in the liberal arts and received his bachelor's degree; soon he got his master's degree, and he was fluent in Latin.

In 1510, he was sent to Rome to settle a dispute within Augustinian monasteries. He was shocked by the lack of prayers of the priests and the luxurious life of cardinals; he started suspecting the efficacy for souls in purgatory.

In 1515, Pope Leon X renewed the sale of plenary indulgence, or paid remission of required punishment for sins. In reaction to this sale of luxury, the elector of Saxony, Frederick the Wise, did not allow the sale of indulgence on his territory. Luther composed a list of ninety-five theses critical of the Church; on the thirty-first of October 1517, he nailed it on the door of the church in the Castle of Wittenberg. Then he reprinted the document and disseminated it throughout Europe.

A few times, Luther was called for trial. At the end of the trial was always execution. The first time, he escaped Rome, and other times, he refused to attend. Finally, Charles V, the Holy Roman emperor, accepted the Nuremberg truce in July 1532. In 1537, Luther was officially excommunicated from the Catholic Church, and he wrote his doctrinal theses, which is called Schmalkalden Articles, to set the Protestant Reformation in motion. That means another religious and political power was established in 1545 under the leadership of Martin Luther.

John Calvin: John Calvin (French: Jean Calvin, or Cauvin), was born in 1509 CE in Noyon, France, and died in 1564 CE in Geneva, Switzerland. Calvin was raised in a loyal Roman Catholic family. His father was a lawyer working for a local bishop and wanted Calvin to become a priest; Calvin was a very devout Catholic. In 1523, he started at the University of Paris to study the priesthood. Later, Calvin decided to study law; between 1528 CE and 1531 CE, Calvin graduated from the law school of Orleans and Bourges. In 1533, he converted to Protestantism, and in 1536, Calvin liberated himself from the Roman Catholic Church.

He decided to leave France permanently, and from 1541 until his death, he settled in Geneva. John Calvin was a leading Protestant Reformer, an essential figure in the second generation of the Protestant Reformation, the founder of Calvinism, and he was a supporter of the renaissance humanist.

Calvinists believe God's predestination or foresightedness of all events perfectly guides those who are destined for salvation. They emphasize the absolute supremacy of God. Calvinism salvation belief is that of destiny (chosen few). Calvinism's core doctrine of predestination, which affirms that God extends grace and gives salvation only to the selected, shows God has discrimination, much like God's chosen people are Jews. Calvin wrote many letters and biblical commentaries, and *Institutes of the Christian Religion*.

Scriptures

There are many denominations in Christianity, and each denomination created their own scripture, but the Bible remains the central part of all scriptures and faiths. The original prints of the twenty-seven books did not exist until 367 CE, written by Athanasius, the bishop of Alexandria, which is the first version of the Bible approved by the First Council of Nicaea in 325 CE.

The Old Testament: The Old Testament is the first part of the Christian Bible and is the original Hebrew Bible. It is the primary text of moral instruction and also history, law, and the prophecy of the people of Israel. Most of the books were written in Hebrew, with some parts in Aramaic. It compromises thirty-nine books, very likely composed between 1200 BC and 100 BC. The Old Testament's first five books are called Pentateuch or Torah, meaning the law as revealed to Moses, and they are Genesis, Exodus, Leviticus, Numbers, and Deuteronomy.

In what language God give Torah to Moses? Hebrew writing started after 1000 BC. It is unknown where the original version of the Torah was saved. It appears the original version never existed and there was no God nor Moses.

The New Testament: The New Testament (Latin: Novum Testamentum) is the final part of the Christian Bible. The original Christian Bible was written in Greek by several authors between 45 CE and 140 CE. The total number of books in the Christin Bible are twenty-seven books, which were collected into a single book over several centuries. They consist of

- the Gospels, explaining the life of Jesus and his teachings;
- Acts, detailing the work of apostles extending the Christian faith;
- the Epistles, teaching the meaning and implication of the faith;
- Revelation, foretelling future events and the apotheosis of the divine intention; and
- letters of Saint Paul and other Christian leaders.

The Divine Comedy: *The Divine Comedy* (Italian: La Divine Commedia) is a long narrative in three parts: Inferno, Purgatorio, and Paradiso. It is an epic poem about the Christian afterlife written by Dante Alighieri around 1308 to 1321. Dante's work is one of the world's most significant works of literature. Fourteenth-century Christians better understood the afterlife from Dante than from biblical teachings.

The Imitation of Christ: *The Imitation of Christ* (Latin: De Imitatione Christi) is a Christian holy handbook. It was written by Thomas A. Kempis around 1418 CE to 1427 CE. *The Imitation of Christ* promotes a life of Christian simplicity. This handbook has been translated into fifty languages and is the second most read book after the Bible.

Denominations

Christianity arose from an established religion of Judaism and soon became the religion of the greatest empire. Christianity was spread all over the Roman Empire, and until 1054 CE, there was one denomination. In 1054 CE, the power struggle escalated when Pope Leo IX excommunicated Michael Cerlarius, the patriarch of Constantinople, the head of the Eastern Church. In the same manner, the patriarch condemned the pope. It is the first time the Christian Church divided into two denominations, Roman Catholic and the Eastern Orthodox.

In 1510 CE, Martin Luther was sent to Rome to settle a dispute within Augustinian monasteries. He was shocked by the lack of prayers of the priests and the luxurious life of cardinals. He became extremely jealous of their life and power, and he started to fight against the pope for energy and money. Finally, in 1545 CE, another denomination and political power was established, the Protestant Reformation.

This process never stopped, and many denominations were created and will be discussed. A few of them are as follows:

Roman Catholicism: The Roman Empire used Christianity for its political goals. Saint Paul the Apostle was a Roman prosecutor and spread the newly raised Christian belief to the Roman Empire. Roman Catholicism, from the very beginning, was instrumental for further development of Christianity, and they developed a very advanced theology and organizational structure, headed by the pope. Presently, the number of Roman Catholics is 1.2 billion worldwide.[104] Catholics attend a weekly Mass based on the Eucharist and serving the body and blood of Christ, consecrated by the priest. Catholics believe in purgatory for cleansing the souls for heaven. In the meantime, the Catholic Church eliminated a lot of scientists who opposed Christian belief, and also encouraged and valued some historical figures, such as Dante, Hayden, and Mozart.

Eastern Orthodoxy: The official name is the Orthodox Catholic Church. Eastern Orthodoxy is the second major Christian denomination, or the first split of the Roman Church. In 1054, during the Great Schism, the power struggle escalated when Pope Leo IX wanted to remove the patriarch of Constantinople. Eastern Orthodoxy adherents live in the Middle East, Greece, some in Eastern Europe, and Russia. Presently, the number of Eastern Orthodoxy members is 370 million worldwide.[105] The adherents follow the faith that was outlined by the first seven ecumenical councils. They do not have papacy other than the interpretation of the scripture.

Sacraments of Roman Catholicism and Eastern Orthodoxy: There are seven frequent sacraments for Roman Catholicism and Eastern Orthodoxy.

1. *Baptism*: In the Catholic Church, infants are washed or sprinkled with water, and the Trinity is invoked to welcome them into the Catholic faith. Also, this frees them from the original sin with which they were born.
2. *The Eucharist*: Also called the Lord's Supper or Holy Communion, the Eucharist commemorates the last supper of Jesus with his disciples. He told his disciples, "The bread is the flesh of my body," and "The wine is my blood."[106] This sacrament is essential and central to Catholicism and is offered in nearly every Mass.
3. *Confirmation*: This is building up what began in Baptism and Holy Communion. An adult or teen must attend catechism classes, after which he or she is confirmed in the faith in a small ceremony.
4. *Penance*: Penance is a sacrament instituted by Jesus Christ, which usually takes place after a private confession in the presence of a priest of grave sins committed after baptism. After the admission, the priest may impose an act of penance and/or say a prayer of forgiveness.

5. *Anointing of the Sick*: This sacrament involves an ordained priest anointing holy oil to a sick or older person, except for those who persevere obstinately to manifest grave sin.
6. *Matrimony*: Christians believe marriage is a gift of God and a divine institution in which Adam and Eve wedded in the Garden of Eden. Marriage can never be broken; even in case of divorce, they are bound to each other until death.
7. *Holy Orders*: The sacrament of Holy Orders is the ordination of ministerial offices, and includes three orders: deacon, priest, and bishop. The priests and deacons ordained by a bishop serve the spiritual needs of others in the church.

Protestantism: The third largest denomination of Christianity came into existence in 1545 and is called the Protestant Reformation. Presently, the number of Protestant Reformation members is 0.72 billion worldwide.[107] The Protestant Reformation was sub-denominated into the following denominations: Pentecostal and Charismatic, Lutheranism, Baptists, Methodist, Reformed and Presbyterians, Amish Mennonites, the Church of Jesus Christ of Latter-Day Saints (Mormons), Adventists or Seventh-day Adventists, Society of Friends or Friends Church or Quakers, Jehovah's Witnesses, Unitarians, Christian Science, and Anglicanism.

Sacraments of Protestants: There are two frequent sacraments for Protestants.

1. *Baptism*: Baptism symbolizes the Protestant entering into God's salvation and covenant. Baptism differs among Protestant denominations. Some include foot washing, while some baptize at different ages.
2. *Eucharist*: Also called the Lord's Supper or Holy Communion, the Eucharist commemorates the last supper of Jesus with his disciples. Protestants have different viewpoints on this, and some observe this sacrament weekly or less.

Pentecostal and Charismatic: The Pentecostal and Worldwide Charismatic movement emerged under the leadership of Charles Fox

Parham, an evangelical Protestant, in the early twentieth century. See chapter 16, "Newer Religions," for details.

Lutheranism: Lutheranism is a sixteenth-century German religious movement under Martin Luther's leadership. The Catholic Church, around 1519, coined him with the term of Lutheran to belittle him. His visit to Rome in 1510 was an eye-opener for Martin Luther; he started suspecting the efficacy for souls in purgatory. Martin Luther was a good politician, and he was able to reduce the authority of the pope and cause substantial financial loss to the Catholic Church and put himself at the head of the new denomination. Presently, the number of Lutheran members is eighty-seven million.

Baptist: Baptists share the fundamental beliefs of Protestantism. The Baptist, once nicknamed the Anabaptist, had its origin in Amsterdam by John Smyth (1554–1612.) Baptist goes back to the Old Testament and believes that baptism has to take place after conversion. Jesus was baptized by the prophet Saint John the Baptist by complete immersion rather than pouring water and sprinkling (Matt. 3:13–17). In the seventeenth century, there were two groups of Baptists. One group believed that Jesus Christ died only for an elect group of people, and those were Calvinists, and another group was the General Baptists. This group believed that Jesus Christ died for everybody.

The Baptist's unity is based on the following convictions:

1. The supreme authority of the Bible in all matters of practice and faith.
2. The believer's baptism.
3. Churches could be composed of believers only.
4. Equality of all Christians in the life of the church.
5. Independence of local church.
6. Separation of church and state.

In the United States, by 1900, the Black Baptists outnumbered Black adherents of all other denominations. The civil rights campaign of the 1960s came out of the Black churches, including Martin

Luther King Jr. Presently, the number of Baptist members is 100 million.[108]

Methodist: The Methodist faith is historically related to the denomination of Protestantism and is an eighteenth-century movement founded by John Wesley. This movement tried, with teachings of brothers John and Charles Wesley, to reform the Church of England from within, known as the Great Awakening. Methodists emphasize the scriptures as an excellent guide to practice. They recognize and practice two sacraments of the Lord's Supper and infant baptism, and hymns are critical to them. Methodism was brought to the USA by Irish immigrants who were converted by John Wesley. In 1735, brothers John and Charles Wesley went to Savannah, Georgia, as missionaries to convert the tribes and establish Methodist churches in the colony. Several organizations affiliated with Methodism include the African Methodist Episcopal Church, the United Methodist Church, the Wesleyan Church, and the Salvation Army. Presently, the number of Methodist members is 75 million.

Reformed and Presbyterians: Reformed and Presbyterians are part of the Protestant denomination who advocate the supremacy of the Bible as a rule of faith and oppose the involvement of the state. The term *Reformed* is regarded as Calvinistic in doctrine and goes back to the sixteenth century. The First Presbyterian Church was organized on a national basis in the sixteenth century in France. Presbyterians believe in the Trinity: God the Father; Jesus Christ, his Son; and the Holy Spirit. Presbyterians emphasize the Bible as an excellent guide to practice faith, and they recognize only two sacraments, the Lord's Supper and infant and adult baptism. Presbyterians acknowledge the universal creeds of the Church. Presently, the number of Reformed and Presbyterian members is 75 million.

Amish Mennonites: Amish Mennonites, Quakers, and Baptists arose from Anabaptists, and their core beliefs are derived from Anabaptists. In 1693, Anabaptists, led by Jacob Ammann, a seventeenth-century elder, were a sect of the radical movement of the Protestant Reformation. The followers of Ammann became known as Amish. They live a simple life, and they do not adopt the conveniences of modern technology. The Amish migration to the USA in the eighteenth century landed in the eastern part of Pennsylvania. In the meantime, the assimilation with Mennonite groups eliminated the Amish from Europe.

Mennonites are named for Menno Simon, a Dutch priest who followed the moderate Anabaptist group. They follow the two sacraments of the Lord's Supper and baptism, as water is a sign of cleansing from sin. Foot washing in the Mennonite church is considered a symbolic practice, and they will do this periodically throughout the year.

The Church of Jesus Christ of Latter-Day Saints (Mormons): The Mormons embrace the concepts of Christianity and the revelation of their founder and prophet, Joseph Smith. Joseph Smith came from a Protestant background, but Mormonism is neither Protestant nor Catholic. Mormons believe in God the Father; his Son, Jesus Christ; and the Holy Spirit as three separate entities united in a single purpose and called the Godhead. People will be punished for their sins, not Adam's sins. Mormons embrace Millennialism, like the rest of Christian denominations, that a Golden Age is coming during which Jesus Christ will reign. Mormons believe in a Second Coming of Jesus Christ, baptism of adults, and prophecy. Mormons do not consume alcohol, tobacco, tea, or coffee. Polygamy was part of Mormon belief, but in 1890 it was renounced. Still, some who broke with the Church have multiple spouses. In Mormonism, kissing before marriage is a sin, and the Church also discourages interracial marriage and dating.

Joseph Smith: Joseph Smith Jr. is the founder and prophet of Mormonism. He was born in the town of Sharon, Vermont, on December 23, 1805. In 1823, at the age of seventeen, Smith claimed that an angel named Moroni, the son of Mormon, came to him and told him, "You had chosen to translate the book of Mormon."[109] The sacred book was written by Mormon, the father of Moroni, about the ancient prophets who inhabited America around the fourth century. Moroni showed Joseph Smith where the inscribed plates of gold were hidden, which covered events from 600 BC to 400 CE. He took Smith at the bottom of a hill in Palmyra, New York, near his house, and Smith dug the ground where the sacred books were buried. Smith transliterated the plates into the Book of Mormon and published it in 1830. Smith was very successful in spreading Mormonism.

Smith, telling his new ideas, was persecuted; he and his brother were jailed for treason. On June 7, 1844, after a short stay in jail, he and his brother were murdered by angry inmates.[110] Presently, the number of Mormon members is fifteen million.

Smith became a prophet in his dream, talking to Moroni. It is similar to Moses speaking to God at the burning bush, and Muhammad speaking to God in a cave.

Adventists or Seventh-day Adventists: Adventists are one of the denominations of Protestant Christian, emphasizing belief in the imminent Second Coming of Jesus Christ. The origin of the Church traces to Battle Creek, Michigan, Unites States, in the mid-nineteenth century. The founder of the denomination, William Miller, was a Baptist preacher. The terms *Adventists* and *Seventh-day Adventists* are used interchangeably. The Seventh-day labeling serves as a reminder of the seventh or the last day of the week. According to the Bible (Gen. 1–2), God created the world in six days, and the seventh day was for his rest. The Jews and Adventists observe the seventh day, or Sabbath, or Saturday, as the rest day.

William Miller: William Miller is the founder of the Adventist movement. He was born in 1782 in Pittsfield, Massachusetts, and died in 1849 in Low Hampton, New York. He was a farmer and an officer in the US Army in the War of 1812. William Miller converted via baptism in 1820 and started to study the Bible, especially the book of Daniel and the Revelation of John. After a long study and analysis of the Bible, he concluded that Jesus Christ would return between March 21, 1843, and March 21, 1844. The Second Coming did not occur. Adventists were disappointed, and William Miller confessed to his mistake and quit from the Adventist movement. After the departure of William Miller, his follower Samuel Snow revised the prediction to a new date of October 22, 1844. A few followers of William Miller continued with this movement and reached the level of prophecy. Presently, the number of Adventist members is seventeen million worldwide.

Society of Friends or Friends Church or Quakers: Quakers is a nickname for the Society of Friends, and the members are called Friends. Quakers, a Christian group, arose from a small group of separatist Puritans in sixteenth-century England, called Seekers, who endorsed Protestant Reformers like Caspar Schwenckfeld, Sebastian Franck, and Dirck Coornhert. The founder of this denomination, George Fox, expressed his thinking to the Friends that there is no need to have a church or middleman between God and humanity. Friends were convinced that their discovery of God would steer to the purification of the entire Christendom. Society of Friends are devoted to living by the Inward Lights, without creed, sacraments, liturgy, clergy, and other churchly forms, and have no need for church garments.

None of the Friends' thinking and ideas were acceptable to Oliver Cromwell and his Puritan government, and George Fox, along with his Friends, were prosecuted with severe punishment. Friends were rapidly growing, and by 1660, there were 20,000 to 60,000 that had been converted from all social classes except the upper level and unskilled laborers. Despite the punishment from 1662 to 1689, the Friends were growing not only in England but also in America, Wales, and Scotland. Until the Toleration Act of 1689, Friends were tormented by penal laws for going to Friends meeting, for not going to the service of the Church of England, and for refusing to tithe. Around 15,000 suffered under those penal laws, and 500 died, but they continued to grow. This religion was not supported by the power of the king, emperor, or government, and, therefore, did not grow in 360 years of existence. Presently, the number of Friends members is 210,000 worldwide.

George Fox: George Fox was born on July 1624, in Drayton-in-the-Clay, Leicestershire, England, and died January 13, 1691, in London. He was from a relatively low-income family and had no formal education. He may have been apprenticed for a cobbler and possibly tended sheep. He became a preacher and missionary and founder of the Friends Church or Quakers. George Fox's attitude toward political and economic conventions was the same as his attitude toward churchly customs.

George Fox was one of the leaders who saw the abuse of the government, Church, and proxies, and had a good intention for English culture. He stood against the mistreatment of people via Church customs and political and economic conventions. Still, the powerful who were in charge did not allow him to bring something better for those who suffered. Up to this date, the Society of Friends are fighting against war, discrimination, poverty, and many other problems hurting humanity.

Jehovah's Witnesses: The Jehovah's Witnesses were founded in Pittsburgh, Pennsylvania, in 1872 by Charles Taze Russel. It is a branch of the Millennialism faction that arose from the Adventist movement; after William Miller's prediction failed, the Adventists' movement divided into several camps. Charles Taze Russel established himself as a controversial and independent Adventist teacher. The Jehovah's Witnesses interpret the Bible word by word and reject the Trinity that denies the divinity of Jesus. They translated the Bible and called it the New World Translation of the Holy Scriptures. They suggested that the coming of Christ's invisible presence signaled the end of the current order of the world and would be followed by his visible presence. They claimed the Millennium would begin after an Armageddon or after a final battle, after which only the Jehovah's Witnesses will rule the world with Jesus Christ in 1914. That prediction never took place. Despite their failure, Russel's teaching encouraged volunteers to circulate his books and pamphlets, including recalculation of the Second Coming. In 1917, Russel's successor, Joseph Franklin Rutherford, reaffirmed its members' belief that Jehovah (Hebrew: Yahweh, God of Moses) is the true God, and the Witnesses are his only chosen followers. The Jehovah's Witnesses have no celebrations, holidays, or festivals except one day, which is the Memorial of Jesus Christ's death at the time of Passover. Every member must spend five hours a week at meetings in a Kingdom Hall, and men are in control in every aspect of life. Presently, the number of Jehovah's Witnesses members is eight million worldwide.

Charles Taze Russel: Charles Taze Russel was born February 16, 1852, in Pittsburgh, Pennsylvania, and died on October 31, 1916, in Pampa, Texas. Russell was trying to move up in controlling others with religion. Russell did not seem to make progress in Presbyterian and Congregational, and he left both to pursue the Adventist movement. After William Miller failed in his Second Coming prediction, Russel saw an opening and started the controversial and independent Adventist campaign. He created the International Bible Students Association and the denomination of Jehovah's Witnesses. He completed additional biblical calculations and preached that Christ's invisible return occurred in 1874 and the final battle was in 1914, followed by a war between capitalism and communism, after which the chosen denomination of Jehovah's Witnesses would rule the earth with Jesus Christ. He wrote many books and other publications.

Unitarians: Unitarians and Universalism are the most undogmatic religious movement of all Christian religious denominations. Unitarianism as an organized religious movement emerged from sixteenth-century Protestant Reformation in Poland, Transylvania, England, and later in America. This religious movement rejected the dogma of the Trinity and emphasized the unity of God. The Unitarian's belief is based on reason, freedom, and in the goodness of human nature. As a result of the encouragement of Ferenc David (1510–1579), a Transylvanian theologian by the name of George Blandrata (1515–1588), an Italian physician and leading Unitarian.[111] Ferenc David taught that the prayers could not be addressed to Jesus Christ. He preached Jesus Christ was only a human, and there was no resurrection. When Catholic Stephen Bathory, a Hungarian endorsed as the prince of Transylvania born in 1533, died in 1586, after the death of John Sigismund, the Unitarians were persecuted. Ferenc David, a Unitarian preacher, was brought to trial in 1579 and condemned to life in prison; within a few months, he died. Presently, the number of Unitarians and Universalist members is eight hundred thousand worldwide.

On the other hand, Unitarians were the most undogmatic religious movement of all Christian religious denominations. They rejected the Trinity, resurrection, and divinity of Jesus Christ, and their beliefs were based on reason, freedom, and in the goodness of human nature. For all the above good intentions, they were not able to expand because of the loss of privileges to the church abusers and politicians. Unitarianism in the United States is evidenced by the four American presidents who were members: John Adams, John Quincy Adams, Millard Fillmore, and Howard Taft.

John Adams was born in 1735 in Braintree, Massachusetts, and died on July 4, 1826, in Quincy, Massachusetts. He was the founding father and the second president of the United States. John Adams was an ideal representative of the American Enlightenment, and his belief as Unitarian or almost nonbeliever made him an exemplary person of the American Enlightenment. John Adams and those who were like him included in the constitution the separation of church and state.

Christian Science: Christian Science is a religious cult founded by Mary Baker Eddy and fifteen of her followers in 1879 in Boston, Massachusetts. Eddy is also the author of *Science and Health with Key to the Scriptures*, which includes the teaching of Christian Science. Eddy emphasizes the elimination of sin and the healing of the sick only through prayer. Christian Science believes in God, who is purposeful and omnipotent, and they accept the authority of the Bible. They believe the crucifixion and resurrection of Christ are essential to the redemption of those who believe in Jesus Christ. Christian Science teaching says that God is a divine mind and is the only real mind. Evil, disease, sin, death, and matter are unreal illusions. Presently, the number of Christian Science members is 270,000. With the progress of science and technology, the numbers are declining.

Mary Baker Eddy: Eddy was born in 1821 in a farmhouse in Bow, New Hampshire, and died in 1910 in Chestnut Hill, Massachusetts. Eddy was brought up in a family that believed in Predestinarian Calvinism. Eddy did not have formal education because of her health and sickness and financial condition. She went through many tragedies in life; because of all the misfortunes and being sick, she was genuinely interested in medicine and the Bible. Eddy's concern was how a God of love could reconcile with the existence of so many miseries and pains. She founded the Christian Science cult in 1879 in Boston in response to the medical system, biblical criticism, Darwinism, and many other secularizing influences that weaken the position of supernatural power and religion.

Anglicanism: The Anglican Church is the established Church of England. This church is one from the sixteenth-century Protestant Reformation and includes both Roman Catholicism and Protestantism. King Henry VIII, born in 1491 and died in 1547, became the Supreme Head of the Church of England in 1534; he broke with Rome because Pope Clement VII would not grant an annulment of his wife Catherine of Aragon so he could marry Anne Boleyn. He reached his goal by having the archbishop of Canterbury, Thomas Cramer, declare the marriage of Catherine of Aragon annulled, and King Henry VIII was free to marry Anne Boleyn. They consider the Bible to be divinely inspired and recognize the Nicene Creed. The Eucharist and baptism are central to worship, as is the Book of Common Prayer.

Beliefs

Christianity has a huge deviation in the practice of the faith around the world, but every Christian in the world believes in God the Father, God the Son (Jesus Christ), and God the Holy Spirit. A few other common beliefs are as follows:

Creation: Creation in Christianity means God created the world from nothing (Latin: Ex nihilo). Before the creation of this world, nothing existed other than God himself. According to the Bible (Gen. 1–2), God created the world in six days, and the seventh day was for his rest. These seven days show God's creation and rest.

1. Light.
2. Sky or heaven.
3. Dryland, trees, and plants.
4. The sun, moon, and stars.
5. Flying creatures and those living in the sea.
6. All the animals on the earth, including man in the image of God.
7. God completed his job and rested.

The age of the earth, according to the Bible (Gen. 1–5), is six thousand years old.[112]

Christianity

Deity: Christians believe that there is only one God, who is the creator and ruler of the world, the source of all moral authority and the Supreme Being. This one supreme God is recognized in the Trinity, which is the Father, the Son (Jesus), and the Holy Spirit.

Zoroastrianism is the oldest monotheistic religion, and it influenced all three Abrahamic faiths. Zoroastrians believe in one Divine Being called Ahura Mazda (the Wise Lord), who is the creator and ruler of the world. It appears that Christianity borrowed the idea of one God from Zoroastrianism, with some modifications.

Ethics: Christian ethics are very well described in the Bible in Colossians 3:1–6 in the New Testament. Christian ethics and morality would be the principles that came from the Christian faith and worship of God. The Ten Commandments are central to both Christian and Jewish ethics and morality.

1. You shall have no other gods before me.
2. You shall not carry the Lord your God's name in vain.
3. You shall not make idols.
4. Remember the Sabbath day to make it holy.
5. Honor your father and mother.
6. You shall not murder.
7. You shall not commit adultery.
8. You shall not steal.
9. You shall not bear false witness against your neighbor.
10. You shall not covet.

Ethics and morality are part of the human struggle from the first time we walked on the earth and will continue regardless of the existence of religion.

Salvation: Salvation, redemption, and deliverance in Christianity mean saving human beings from suffering. John 14:6 describes very well that sinful human beings will never get to heaven on their own. This doctrine is known as original sin; for this reason, Jesus was born, died, and was resurrected. In Genesis 2:7, Adam was created by God from the dust of the ground (clay). God breathed into his nostrils the puff of life, and then Adam turned to a living creature. In Genesis

2–15, God placed Adam in the Garden of Eden to tend and watch over it, and God warned him, you are allowed to eat every fruit in the garden, avoid the tree of knowledge of good and evil. If you consume its fruit, you are sure you die. In Genesis 2:21-25, God created a woman from the rib of Adam and brought her to Adam. In Genesis 3:1-24, the serpent or Satan encouraged to woman to eat the fruit from the tree of knowledge of good and evil. She took the fruit and ate it, and gave some to her husband. Upon eating the fruit, both realized they were naked and used leaves to cover themselves. God cursed them for their disobedience, man with a life of hard labor, and woman the pain of childbirth, and cast them out of the Garden of Eden.

On the other hand, salvation is a fascinating myth. The death of Jesus was not a voluntary action by Jesus. He was accused of treason and arrested. After the court judgment, he was executed by order of Pontius Pilate, one of the most potent Roman prefects, or governors, of Judaea, under the emperor Tiberius. Jesus was beaten until his skin was peeled off him, then crucified by being nailed to a cross he was forced to carry to Calvary (the hill on which he and two criminals were crucified). Jesus was first to be killed. His head was thrown out, and then he was nailed on the cross on top of a hill; the strong Roman army was watching the executed person until it turned to clean bones, and the army discarded them. Like other religions, Christianity had to embed fear in followers' minds to control and subdue them. Jesus created fear in his followers' minds, according to John. In John 14:6, Jesus told him, "I am the way, the truth, and the life. No one can come to the Father except through me."

Death and the afterlife: Christians believe death is not the end of life because Jesus promised and taught that those who believe in him will be given eternal life in heaven. Judgment will happen after death. They will come into the presence of God and will be judged for good and evil during their lifetime. Those who did good will go to heaven, and those who did bad go to hell of eternal fire. Some Christians believe in a two-stage judgment after death and at the end of time.

The Roman Catholic Church teaches that after death, there is a state of purgatory. It is a place where the deceased family members pay a large sum of money, and the sinner will be purified, and the dead is entitled to heaven.

End-time or the end of the world: The end-time is coming (from eschatology, Greek: *eschatos* means last) is a branch of theology apprehensive of the final destiny of humanity. Jesus will come back at the end-time, known as the Second Coming or the Rapture. During the Rapture, Jesus Christ carries Christians, dead and alive, to heaven to meet the Father (1 Thess. 4:17) and defeat the most prominent opponent of God, Satan.

Rituals and Customs

Christian rituals, customs, and worship have a variety of methods, depending on the denomination.

Baptism: Christianity adopted Baptism from Judaism into the life of the Church, and is performed by a member of the clergy. Each denomination has differing views and beliefs about baptism.

Eucharist: The Eucharist is also called Holy Communion or the Lord's Supper. This ritual commemorates the last supper of Jesus with his disciples. He told his disciples, "The bread is the flesh of my body." and "The wine is my blood."[113] Some denominations use baked bread and wine and others use unleavened bread and grape juice. Eastern Orthodoxy and Roman Catholicism have a similar Eucharistic belief. The only difference is the Roman Catholics consider the Eucharist a sacrament. Eastern theology thinks the invocation of the Holy Spirit is part of the Eucharist.

On the other hand, Jesus was captured by the strongest empire of the world. Nobody could see or visited Jesus. After a quick and straightforward trial, he was executed and nailed on the cross for public view, unlike the authority that could allow him to have a last dinner with his followers.

Prayer: Prayer is a means of communication that allows Christians to talk to God, which includes praising God and asking for help. The model of prayers is abundant in the Bible. Prayer can be performed in different ways, and every denomination has its way of praying.

Religious services: Religious services and religious gatherings take place in a house of worship or a church every week, usually on Sundays.

Mostly these gatherings consist of prayers and sermons, studying and reading of scriptures, singing, and participation in rites like Communion.

Confirmation: Confirmation is a ritual practiced by some denominations, like Roman Catholic, Orthodox, Anglican, and Lutheran Churches. It usually takes place in young adulthood. Confirmation is usually preceded by instruction in catechism. Most Protestants disagree with confirmation as a ritual or sacrament. Instead, they accept approval for baptized members into full membership of the church.

Marriage rituals: Christians believe marriage is a gift of God and a divine institution. Marriage can never be broken; even in case of divorce, they're bound to each other. In the old days, the wedding was straightforward. The service was written into the book of common prayers, and the couple could promise themselves to each other in any place or time. Now a Christian wedding is more complicated; it takes place in a church, with a priest guiding the couple in the exchange of vows. After the service, a reception takes place in a separate location.

Sin and repentance: Christians believe human beings are born with sin. The origin of the sin attributes to Adam, who disobeyed God in eating the forbidden fruit of the tree of knowledge of good and evil. Adam transmitted his sin and guilt by heredity to his descendants. Before Jesus, every human being was separated from God in eternal torment and spent time in hell after death. Jesus took the punishment for his faithful Christians by dying on the cross. Also, Christians who violate the Ten Commandments, Jesus's teachings, causing harm to others, and committing sin will distance them from God. If Christians recognize themselves as sinners, then they have to confess their sins to God and ask for forgiveness. The sinners are expected to repent and turn their backs on their sin and come around to seek God, and Jesus will forgive them. Christians practice repentance regularly, which is scheduled in many churches' liturgies.

Festivals and Holy Days

Ascension Day: Ascension Day notes the belief that Jesus ascended to heaven after he was resurrected on the fortieth day of Easter. It is one of the essential feasts in Christian churches.

Ash Wednesday: Ash Wednesday is the first day of Lent, which is celebrated forty-six days before Easter Sunday, mostly in February or March. On this day, the ashes of the burnt palms from the previous year's Palm Sunday are placed in the shape of a cross on the foreheads of the faithful. It reminds them of how sorry they are for their sins and to change their behavior. On this day, the faithful also fast (abstain from eating).

Lent: Lent is a period of forty days starting on Ash Wednesday and ending on Easter, and Christians spend this time preparing for Easter. They concentrate on spiritual studies, prayer, self-denial, and even fasting. In some denominations, those over the age of fourteen are to refrain from eating meat.

Palm Sunday: Palm Sunday is observed a week before Easter. This day is a commemoration day of the Gospel story of Jesus's triumphal entry into Jerusalem, riding upon a donkey to show his humility. His followers scattered palms in his path and called him Hosanna or Savior.

Maundy Thursday: Maundy Thursday is a ceremony of washing the feet of the poor, and it is the Christian holy day coming on the Thursday before Easter. This day commemorates the Last Supper of Jesus Christ with the apostles.

Easter: Easter is the first Sunday after the first full moon in March or April, and is the chief Christian feast. It is a joyful day for Christians to celebrate the resurrection of Jesus Christ. A resurrection is a massive event that Christians believe that Jesus Christ had the authority to die for the original sin of the Christians who believe in Jesus. Orthodox Easter comes a couple of weeks after Easter for the rest of the world's Christians.

Pentecost: Pentecost is the Sunday occurring fifty days after Easter. The Pentecost is a Christian festival celebrating the descent of the Holy Spirit on the disciples of Jesus after Jesus's ascension; it coincides with the Jewish festival of Shavuot. This Holy Spirit descent on the disciples was experienced like tongues of fire and strong wind. It gave them strength for speaking about Jesus's life, death, and resurrection to the people of different languages.

Assumption of the Virgin Mary: Assumption Day is short for the Assumption of the Virgin Mary, and one of the chief feasts admiring Jesus's mother, Mary. The Catholic Church, Eastern, and Oriental Orthodoxy believe that, at the end of her life, the Virgin Mary was physically taken into heaven.

All Saints' Day: All Saints' Day, or All Hallows' Day or Hallowmas or the Feast of All Saints is a feast day celebrated on November 1. On this day, Christians honor all the saints of the churches, known and unknown, who have attained heaven. The preceding evening, called Halloween, means All Hallows' Eve.

All Souls' Day: All Souls' Day is the feast of the Catholic Church, Eastern Orthodox Church, and a few other Christian denominations, and is celebrated on November 2. The churches pray for the souls of departed Christians still suffering in purgatory.

Advent: The Advent is a season or a period marked by the four Sundays before Christmas. During the Advent season, Christians prepare for the coming of Jesus Christ to celebrate his birth.

Christmas: Christmas is the feast of the birth of Jesus, celebrated all over the world on December 25. Easter Orthodox Christians celebrate the birth of Jesus on January 7. Jesus's birth is narrated in the Bible, in the New Testament Gospels of Luke and Matthew. There is a disagreement between Christians because some regard Jesus's birth as described in the Bible theological truth, but not historical.

Influence of Zoroastrianism on Christian Faith

Zoroastrianism was the world's oldest monotheistic religion of the ancient Aryans, the inhabitants of Khorasan. Based on archaeological evidence and linguistic comparison, modern scholars believe that Zoroaster must have lived sometime between 1500 and 1200 BC. Cyrus the Great, around 550 BC, founded the Persian Empire (the largest empire in history) with the official religion of Zoroastrianism. At this time, the Avestan manuscript of Zoroastrianism was saved at the Royal Library at Ishtakhr.

With the existence of Zoroastrian to the east for so long, and especially with occupying forces of the Persian Empire, there is no doubt that the Zoroastrianism impact made itself known in many ways, two of which are by loaning words and providing concepts to the integration of religious ideas and mythic details.

Paradise: In the Indo-Aryan language, Avestan, paradise is called *pairidaeza*, which is composed of two words, *pairi*, which means around, and *daeza*, which means wall, and pairidaeza means "a wall enclosing a garden." Jews borrowed paradise beginning 500 BC, when Yehud, the Jewish province, was a Persian province. In Genesis, paradise, or Greek paradeisos, is used to translate to the Garden of Eden.

Heaven and hell: In Zoroastrianism, heaven is called the House of Song, filled with pleasure and happiness; hell is called the hellish abyss and is filled with discomfort and darkness. The Hebrew Bible had no concept of heaven and hell until after the liberation of the Jews, around 500 BC, when Jews included heaven and hell in the Hebrew Bible with the same idea as Zoroastrians.

End-times or the end of the world: Zoroastrianism believes in the final judgment in the Avestan term *Frashokereti* (making wonderful) at the end of time. Before the liberation of Jews, the Hebrew Bible had no concept of Day of Judgment; after around 500 BC, Jews included the Day of Judgment in the Hebrew Bible, with the same idea as Zoroastrians. Christianity went further and borrowed the coming of a final messiah.

Resurrection: Zoroastrians believe all souls are resurrected and reunited with the resurrected earthly bodies. The Hebrew Bible did not have any concept of resurrection until after 500 BC, which Jews added to the Hebrew Bible.

Messianic prophecy: Zoroastrians believe in the end of time with the arrival of a final messiah or Saoshyant (Avestan: One Who Brings Benefit). The Hebrew Bible had nothing about Messianic prophecy or Messiah. After around 500 BC, the Jews included the concept of Messianic prophecy in the Hebrew Bible, similar to Zoroastrianism.

Death and the afterlife: Christians borrowed the final judgment from Zoroastrians and modified eternal hell or heaven. In Zoroastrianism, those who follow falsehood serve in hell, and at judgment day, everybody lives in paradise.

Free will: Ahura Mazda (Wise God) gave humans free will and made him or her morally responsible for his or her actions. Christianity, borrowed the concept of free will and adapted it to their holy scriptures.

Creation: Ahura Mazda (Zoroastrianism) decided to create different creations as follows:

1. Created sky, shining and bright
2. Created pure water
3. Created sun, moon, stars, and earth, round and flat with no mountains and valleys
4. Created plants, sweet without bark and thorns
5. Created animals big and small
6. Created the first man called Gayōmart from clay
7. Created fire and distributed it among the whole creation

It appears the Bible borrowed the story of creation from Zoroastrianism and modified it accordingly.

Morality

Morality has nothing to do with belief in God and religion. Morality is hardwired in the human brain that makes the distinction between right and wrong, and morality is an evolved faculty with a genetic basis.

It is clear from observation of the Golden Rule of various religions that the moral codes that are working to minimize the suffering of others arise in all faiths. Among all religions, Jainism, which does not believe in God, ranks the highest in morality.[114]

Let's review the universal moral codes independent of the existence of God, which is called the Golden Rule of various religions, in chronological order.

- Hinduism: The oldest organized religion is Hinduism, which goes back to the Indus Valley civilization (2300–2000 BC). In Mahabharata (Sanskrit: Great Epics of the

Christianity

Bharata Dynasty) book 13, section CXIII, verse 8 says, "One should never do that to another which regards as harmful to one's self, and this called the rule of dharma. Other behavior is due to selfish desires."

- Zoroastrianism: Zoroastrianism is the world's oldest monotheistic religion, and its founder was Zoroaster (1500–1200 BC). Zoroaster says, "Do not do unto others whatever is injurious to yourself."
- Judaism: Judaism is the oldest Abrahamic religion. In the Old Testament, Leviticus 19:18 says, "Do not explore revenge or bear resentment against a fellow Israelite, but make sure love your neighbor as yourself."
- Jainism: The founder of Jainism is not known. Mahavira (599 BC) is the most regarded Jain, who gives Jainism its present-day form. They avoid the suffering of any living thing, and they respect the environment and do not believe in God. The sacred scripture Agamas says, "Just as sorrow or pain is not desirable to you, so it is to all who breathe, exist, live or have any quintessence of life. To you and all, it is undesirable and repugnant."
- Buddhism: Buddha was born in Nepal in 563 BC. Buddha said, "One who, while himself seeking happiness, oppresses with violence other beings who also desire happiness, will not attain happiness hereafter" (Dhammapada 10). He also said, "Hurt not others in ways that you would find yourself hurtful" (Udanavarga 5:18).
- Taoism: The founder of Taoism, Lao Tzu, was born in 601 BC in China. "The sage has no interest of his own, but take the interests of the people as his own. He is kind to the kind; he is also kind to unkind: for virtue is kind. He is faithful to the faithful; he is also faithful to the unfaithful: for virtue is faithful" (Tao Te Ching, chapter 49).
- Confucianism: Confucius was born in Shandong in 551 BC. He did not believe in God, and he was a teacher. He said, "Do not impose on others what you would not choose for yourself."

- Christianity: The Christian Bible has many instances recommending the Golden Rule. Here are just a few. "Do to others whatever you would like them to do you" (Matt. 7:12). "Do to others as you would like them to do to you" (Luke 6:31). Also, "Do not seek revenge or bear a grudge against a fellow Israelite, but love your neighbor as yourself" (Lev. 19:18).
- Islam: The Arabs considered the survival of tribes paramount, and the ancient rite of blood vengeance ensured them. Muhammad was from the same culture as the survival of the tribes. During the rise of Islam, Muhammad and his followers survived through bounty raids and looting of caravans. In the Koran, there are several moral guidance and religious laws, but there is a lack of the Golden Rule. But in Hadith, a Bedouin came to the prophet, grabbed the stirrup of his camel, and said, "O the messenger of God! Teach me something to go to heaven with it." The prophet said, "As you would have people do to you, do to them; and what you dislike to be done to you, don't do to them. Now let the stirrup go! [This maxim is enough for you; go and act by it!]" (Kitab al-Kafi, vol. 2, p. 146).
- Sikhism: The Sikh religion and philosophy sprang from Hinduism and Islam. The Guru Granth Sahib says, "If thou desirest thy Beloved, then hurt thou, not anyone's heart" (Guru Arjan Dev Ji 259, Guru Granth Sahib).
- Bahaism: The Tablets of Baha-Ullah say, "Blessed is he who preferreth his brother before himself" (Baha-Ullah)

Summary

Christianity sprang from Judaism and is one of the Abrahamic religions. The beginning of Christianity starts with the magical birth of Jesus and his adult life, teaching, death, and resurrection in a historic city of lower Galilee, in northern Israel, during the height of the Roman Empire. Those Jews and non-Jews who called themselves the follower of Jesus designated a new religious community that was first called Christ (Greek: Christos) in the Syrian town of Antioch, possibly 35 to 40 years after the death of Jesus. The Christian faith has become one of the largest religions in the world; there are 2.3 billion Christians around the world.

When then the Romans conquered another country, they usually let them preserve their culture and religion. Sometimes, Romans adopted other religious gods and included them in their groups of gods. But Romans never came along with Jews and their faith. They could not adapt Judaism as their official religion of the Roman Empire because they could not have succeeded in their political strategies with the Jewish doctrine of God's people and the promised land.

Soon after the Jewish rebellion against the Roman Empire that began in 66 CE, the Roman Empire focused on the temple and its destruction in 70 CE. In this war, a hundred thousand Jews were killed, and ninety thousand enslaved. The Roman Empire was trying to create another religion parallel to Judaism. Peter/Paul of Tarsus (Saint Paul) was a Roman citizen and a Roman prosecutor. After his conversion on the road to Damascus, he was a very passionate missionary in spreading Christianity in the empire.

Finally, Constantine I (Latin: Flavius Valerius Constantinus) learned from his predecessors that bearing the banner of Christianity would be to his political advantage, especially during the Battle of Milvian Bridge outside Rome in 312 CE. Constantine called Church officials to the council of Nicaea in 325 CE to eliminate the conflict within the Church. He was the first speaker of the council (even while he had not converted yet). Finally, after many years of using Christianity for his political advantage, he did convert to Christianity on his deathbed in 337 CE to help the future empire. In 380 CE, Emperor Theodosius I adapted Nicaea, and in 325 CE, Christianity became an official religion of the Roman Empire.

The Christian Bible consists of the Old Testament and the New Testament. The New Testament is the final part of the Christian Bible. The total number of books in the Christian Bible is twenty-seven books. The original prints of the twenty-seven books did not exist until 367 CE; they were written by Athanasius, the bishop of Alexandria, which is the first real version of the Bible approved by the First Council of Nicaea in 325 CE.

Regardless of how religion is progressive, the primitive faith always has some roots in it; in Christianity, it is the figure of Jesus Christ on the cross and statues of Mary and other saints.

Let's see if the element of fear exists in Christianity. Fear is a tool to keep people under tight control by leaders and religious authorities. In Christianity, there are two primary components of the fear and reward after death:

1. *Heaven* is the abode of God, angels, and souls of the righteous after death, filled with singing praises to God and heavenly banquets.
2. *Hell* is a place after death in which the evil person is subject to punishment and burning and torment for eternity.

CHAPTER 13

Islam

The Banu Qaynuqa in 624 CE, one of the richest tribes of the three Jewish tribes in Medina who helped Muhammad in Medina in 622 CE with food, clothing, and place of living as refugees. Later in 624 CE, Muhammad and his followers by force besieged the Banu Qaynuqa and directed them to leave their homeland and leave their properties to Muslims (Copyright Alamy Stock photo and Copyright iStock credit: Pe3check).

> Dance, when you're broken open. Dance, if you've torn the bandage off. Dance in the middle of the fighting. Dance in your blood. Dance when you're perfectly free.
>
> —Rumi

Islam is one of the significant Abrahamic (Arabic: Ibrahim) monotheistic religions of Middle Eastern culture. Islam was founded by the Prophet Muhammad in the Arabian Peninsula in the seventh century CE. In Arabic, *Islam* means submission or surrender. The believer of Islam is called Muslim, who accepts submission to the will of God (Arabic: Allah). God is viewed as the only one God, who is creator, restorer, and sustainer of the world, and he is omnipotent, omniscient, omnipresent, and omnibenevolent.

In the Islamic religion, Muhammad is the last in succession of 124,000 prophets. Twenty-five prophets are named in the Koran, including Adam, Noah (Arabic: Nuh), Abraham (Arabic: Ibrahim), Moses (Arabic: Musa), Solomon (Arabic: Sulayman), and Jesus (Arabic: Yasue or Messiah). The 124,000 prophets were sent by God for the guidance of the Middle East only, and the names of those prophets are from Abrahamic religions, which is not in the Bible.

Islam's sacred scripture, the Koran, notes the will of God to which Muslims must surrender. The Koran was revealed through God's messenger, Muhammad, in 610 CE by the archangel Gabriel (Arabic: Jibril). The revelations continued until Muhammad's death on June 8, 632 CE.

The religion taught by Muhammad and prophecy was first adopted in a family circle of his wife, Khadija, his ten-year-old cousin, Ali ibn Abi Talib, adopted son, Zaid, and close friend Abu Bakr. Muhammad started to preach in public in 613 CE (Koran 26:214). In 619 CE, after the death of Prophet Muhammad's uncle Abu Talib, his other uncle, Abu Lahib, became head of the Hashim clan. City leaders became hostile to Muhammad. Abu Lahib did not protect Prophet Muhammad, and the persecution of Prophet Muhammad increased. In 622 CE, Prophet Muhammad was invited by Arabs and Jews of Yathrib, later called Medina. In the same year, on September 24, 622, Prophet Muhammad secretly slipped away to Medina, where the history of Islam started.

After the complete defeat of the Arabs and Jews in Medina, Prophet Muhammad created a robust, organized community of military forces. In January of 630, Muhammad, with ten thousand armed men, headed to Mecca and captured it in a short time. Prophet Muhammad was an excellent politician and religious leader, as history bears witness. He was an effective ruler of Mecca and Medina. His teaching was rapidly growing, and the Islamic empire became the largest empire of the world; at present, Islam is the second largest religion, with more than 1.8 billion followers.[115]

Ancient life on the Arabian Peninsula: Nomadic Bedouin tribes dominated the Arabian Peninsula. Their tribe was composed of a couple of clans. The system of society was patriarchal. The name and inheritance went through the male lineage. The tribe provided every means of living and protection for its members. The Nomadic Bedouin tribes' main occupation for centuries was pastoralists. They were dependent on their small herds of sheep, camels, horses, and goats for milk, meat, cheese, wool, fur, and food substances. Their climate and environment were harsh. About 80 percent of their land was barren desert, with almost no source of water. Yatha' Amar Watar I rebuilt the ancient earthen Ma'rib Dam in 750 BC in the southern Arabian Peninsula city of Yemen for agricultural purposes. This structure created a vast agricultural field. The Ma'rib Dam collapsed in 570 CE and brought down the ancient commercial empire.

As a result of this dam collapse, thousands of people migrated to other parts of the peninsula. In the Arabian Peninsula, there are only two other places, like Ta'if, a city in Mecca, and Yathrib, or Medina, where agricultural communities could flourish.

The Arabian Peninsula was invaded in 540 BC by the Persian Empire and in 330 BC by the Greek Empire and later by two major empires, Roman and Persian, until the rise of Islam. Both empires were wrestling with how to control the vast territory and fragmented populations, whether by force or injecting religious belief. They concluded religious belief would be the better way. That strategy resulted in defeating both empires after the rise of Islam. The

Arabian Peninsula's religion was predominantly polytheistic and had also Judaism, Christianity, and Zoroastrianism, especially in Mecca and Medina.

At the end of the fifth century AD, merchants found Bedouin tribes suited for transporting goods and as guide caravans across the barren deserts because of familiarity with water sources. Through this path, goods from East Africa, India, and Yemen traveled on to both the Persian and the Byzantine Empires.

So, lack of water sources and pastureland created a rough tribal culture of continual bounty raids and looting (ghazu) of other tribes. Tribes were enemies to one another and a tribe's goats, sheep, camels, horses, slaves, women, and children were under permanent threat. These attacks were at the peak when drought caused groundwater to be scarce.

Leaders

Muhammad: Muhammad is the founder and prophet of Islam. His full name is Abu al-Qasim Muhammad ibn Abd Allah ibn Abd al-Muttalib Ibn Hashim. Muhammad's family was of the Hashim clan of the Kuraish or Quraysh tribe. Muhammad was born in 570 CE in Mecca, and died on June 8, 632, in Medina. In 570, the year Muhammad was born, South Arabian king Abraha al-Ashram decided to invade Mecca, but this invasion was foiled by God's intervention (Koran 105:1–5). God the omnipotent, omniscient, omnipresent, and omnibenevolent acted very violently and killed a vast population of innocent soldiers. God is the creator of the world; he could change the heart and mind of King Abraha without killing so many with violent action.

The story of God's intervention is only in the Koran. History supports that Muhammad's father, Abd Allah ibn Abd al-Muttalib, died before Muhammad's birth; his paternal father, Abd al-Muttalib, the chief of Hashim clan, took care of him. His mother, Aminah, died when he was six years old, and his grandfather died when he was eight years old.

After the death of his grandfather, his uncle Abu Talib, who was the head of the Hashim clan, took care of him. At the age of twelve, Muhammad traveled to Syria with his uncle Abu Talib until the age of twenty-five.

At the age of twenty-five, Muhammad was employed by a wealthy middle-aged widow named Khadijah (or Khadija bint Khuwaylid), the daughter of Khuwaylid ibn Asad, a leader of the Quraysh tribe in Mecca. Khadija was born in the sixth century CE and died in November 619 CE. Muhammad's job was to supervise the transportation of Khadija's commodities to Syria. Muhammad influenced Khadija so much that she offered to marry him; at the time of their marriage, she was forty years old. She gave birth to four daughters and two sons, who died young, and he had another son by his concubine Maria, who also died young.

Slavery was common on the Arabian Peninsula; young Zayd ibn Haritha was kidnapped by horsemen from the Qary tribe and was sold to a merchant from Mecca Hakim ibn Hizam as a slave for four hundred dinars. Hakim ibn Hizam gave the boy as a gift to his aunt Khadija bint Khuwaylid. The boy remained with Khadija until she married Muhammad. Muhammad was very attached to Zayd, and he adopted him as his son. Zayd's father and uncle found out about him and approached Muhammad and offered a ransom for the return of their child. Muhammad agreed that Zayd should be allowed to select his future, and he would be released without ransom. He called Zayd in front of a group; Zayd expressed that he wanted to stay with Muhammad. Muhammad took Zayd to the steps of the Kaaba and announced to the group of people that Zayd had become his son with rights of inheritance. Zayd was named Zayd ibn Muhammad after this event, and he legally became Muhammad's son. Zayd ibn Muhammad was the third person to become Muslim after Muhammad's wife Khadija and Muhammad's cousin Ali ibn Abu Talib.

Fatimah, Muhammad's daughter, who married Ali ibn Abi Talib, is regarded by Shiite Muslims as the divinely ordained successor of Muhammad. After the death of Khadija in November 619 CE, Muhammad did not marry until his arrival in Medina in September

622 CE. In Medina, Muhammad mediated between two rival clans of Banu Khazraj and Banu Aws. Besides that, he created a covenant between all tribes in Medina for cooperation and support of each other. Muhammad became a great leader and was expected to have a large harem, since polygamy was the culture of Arabian society. Muhammad was on the rise, and the first khalifah of Islam, Abu Bakr, and the second khalifah of Islam, Umar ibn al-Khattab, offered their daughters to Muhammad.

Aisha bint Abu Bakr, based on some tradition of hadith, was at the age of six or seven (some other sources say she was nine years old) when she married Prophet Muhammad in a small marriage ceremony.

According to Bukhari, Ibn Ishaq, Ibn Hisham, Al-Tabari, Ibn Sa'd, Majlisi, and Qayyim, Muhammad married nineteen wives, had fourteen engagements and broken contract, and ten refused proposals from the girls' family. Those Muhammad married are as follows: (1) Khadija bint Khuwaylid, (2) Sawda bint Zam'a, (3) Aisha bint Abu Bakr, (4) Hafsa bint Umar, (5) Zaynab bint Khuzayma, (6) Hind (Umm Salama) bint Abi Umayya, (7) Zaynab bint Jahsh (see following paragraph about Zaynab's faith), (8) Rayhana bint Zayd ibn Ami, (9) Juwayriyya bint al-Harith, (10) Ramlah (Umm Habiba) bint Abi Sufyan, (11) Safiyya bint Huyayy ibn Akhtab, (12) Maymuna bint al-Harith, (13) Mariyah or Maria bint Shamoon ai-Quptiya, (14) Mulayka bint Kaab, (15) Fatima al-Aliya bint Zabyan al-Dahhak, (16) Asma bint Al-Numan, (17) Ai-Jariya, Amran bint Yazid, and (18) Tukana al-Quraziya.

In 625 CE, Prophet Muhammad proposed to Zaynab bint Jahsh, his cousin, to marry his adopted son Zayd ibn Muhammad. Soon, his son and Zaynab married. The following year, Muhammad stopped at his son's house. This time when Muhammad saw his daughter-in-law, God created some special feeling in Muhammad's heart about Zaynab. Muhammad told Zaynab and left her house. Zaynab told Zayd what Muhammad told her. By the end of 626 CE, Zayd had to divorce Zaynab, and soon after that, Muhammad married Zaynab. Muhammad realized that his action was immoral, so he issued a couple of revelations to calm his followers (Koran 33:4–5; 33:37, 40; 66:1.)

When Muhammad was forty years old, he was meditating in a cave in Mount Hira outside the city of Mecca. The archangel Gabriel (Arabic: Jibril) came to him and said, "Recite," and three times, Muhammad refused. Finally, the archangel Gabriel said, "Recite thou in the name of thy Lord who created; Created man from clots of blood" (Koran 96:1–18). The statements given by the archangel Gabriel to Muhammad proclaimed the oneness and power of God, to whom worship should be made, and face judgment.

After the first revelation, Muhammad was concerned and unsettled, but Khadija reassured him of his prophecy. Muhammad was illiterate, and the Koran called Muhammad the "unlettered prophet" (Koran 29:48).

Muhammad continuously received revelation. For the first three years of his prophecy, he was speaking in private with his wife, his cousin, and his wife's cousin. Finally, God commanded him to move on to public preaching (Koran 15:95). The doctrine of Muhammad, who attacked the religious beliefs and practices of the Kuraish or Quraysh tribe, was not acceptable to Meccans. The Meccans refused to trade or intermarry with the Hashim clan. Therefore, his uncle Abu Lahib, who was chief of the Hashim clan, did not agree with Muhammad's doctrine, and he did not protect Muhammad. Instead of changing the heart and mind of Meccans and his uncle Abu Lahab, God sent a revelation about Abu Lahab (Koran 111:1–5).

On February 26, 621 CE, Muhammad was miraculously transported to Jerusalem to pray with Abraham, Moses, and Jesus by archangel Gabriel (Arabic: Jibril). This event, called Night Journey, came while Muhammad was sleeping in the Cave Hira (some say in the Al-Haram Mosque). After prayer in Jerusalem, Gabriel ascended Muhammad to heaven, where he met with God, and God imposed on him the five daily prayers.

In 622 CE, Prophet Muhammad was invited by Arabs and Jews of Medina. In the same year, on September 24, 622, Muhammad secretly slipped away to Medina because the angel Gabriel told him of a planned assassination by the Quraysh. Muhammad chose a different route to Medina, and he entered Medina, where the history of Islam started.

When Muhammad settled in Medina, first, he built a simple mosque (Arabic: masjid). Tree trunks supported the roof, and a stone marked the mihrab or qiblah (direction to pray from Mecca to Jerusalem). This mosque included a courtyard for religious, social, and political concerns of the Muslims. At the beginning, Arabs, Christians, and Jews were providing living places, food, and clothing for the immigrants or refugees (Arabic: Almuhajirin) from Mecca or Ummah. Jews were very friendly and enhanced Muhammad's knowledge with Jewish scripture. So, for all the kindness of the Jews, Muhammad made a few new changes in Islamic practices to align closely with Judaism. Those changes were the following:

1. Communal prayers on Friday, while Jews prepare for Saturday
2. Praying toward Jerusalem like Jews

Gradually, the relations between Jews and Muhammad deteriorated. Jews did not like Muhammad's prophecy because the prophecy era was over for Jews based on their myths and beliefs, and they also resented Muhammad's claim of the Night Journey. Most importantly, the financial condition of Muslims in Medina was a significant point of conflict; all Muslims found themselves out of work. Jews owned the land in Medina as the residents of Medina. Muslims could not perform agriculture to support their families and communities.

Muhammad and his followers turned to the longstanding Arabian culture of the ghazu, or bounty raid. Muhammad himself was as a leader, the initiator, organizer, and participant of the bounty raid and looting of caravans. The Muslims' income was coming from bounty raids of the Meccans' caravans to Syria or from Syria to Mecca.

The Prophet Muhammad's bounty raids and looting were unsuccessful because merchants rerouted their caravans. Poor and desperate, Meccans were flocking to Medina to make easy money, joining the Muslims. During these conflicts between Muslims and Jews, Muhammad changed the direction of Muslims' prayers in 630 CE, from Jerusalem to the Meccan Kabah (Koran 2:144).

Battle of Badr: In 624 CE, after two years of Muhammad's bounty raids on Meccan caravans, Meccan tribes mobilized an army to Medina under Abu Sufyan to confront Muhammad and his followers. Muhammad prepared a raiding army of about three hundred Muslims, led by Muhammad himself. Gabriel helped him at the time of crises; thousands of angels descended from heaven to aid Muhammad. The Meccan army was defeated, and the entire army slaughtered by the order of Muhammad. He scored a complete victory.

Muhammad got enormous prestige. Meccans tried again in 625 CE in the Battle of Uhud and were back in 627 CE in the Battle of the Trench to capture Medina, but both attempts were unsuccessful.

After these two battles, Muhammad assured his followers in the next fight for ownership of land and wealth. After the Meccan defeat, Muhammad was emboldened to exterminate Jewish tribes to occupy all of Medina and establish the Islamic state.

After the Battle of Badr in 624 CE, angels told Muhammad to attack the Jewish tribes of Banu Qurayzah and Banu Qaynuqa. The Banu Qaynuqa tribe was the wealthiest tribe in Medina. They did not own farmlands; they were traders, arms manufacturers, jewelry makers, and owned a marketplace. They lived in fortresses and beautiful houses. In April 624, Muhammad laid siege to the Banu Qaynuqa quarter. After fifteen days, the Banu Qaynuqa tribe was compelled to surrender.

Muhammad gave them three days to leave, with all their belongings, businesses, fortresses, and marketplace to be submitted to Muslims. Muhammad divided the Banu Qaynuqa tribe's booty to the Ummah and kept one-fifth for himself.

After the Battle of the Trench in 627 CE, and success of Muhammad's Quraysh forces, Muhammad charged the Banu Qurayzah with treason, and they were besieged by the Ummah under the commandment of Muhammad. Ultimately the Banu Qurayzah tribe surrendered, and their males were savagely slaughtered at the command of Muhammad. Muhammad divided the Banu Qurayzah tribe's booty to the Ummah, including the enslaved women and children, and kept a more significant portion for himself.

In or around July 628 CE, Muslims besieged the Banu Nadir tribe for a couple of days, under the commandership of Ali Ibn

Abu Talib. During this siege, a Jew called Azwak attempted to kill Muhammad, but Azwak was killed by Ali Ibn Abu Talib. After this event, the Jews surrendered and agreed to leave Medina with only one load of a camel, including their property, except silver, gold, and army equipment. Safiyya, the daughter of Huyayy ibn Akhtab, had to marry Muhammad after her husband, Kinana, the chief of the Jews in Khyber, was savagely tortured and beheaded under the supervision of Muhammad. The property Muslims received did not count as booty, and all was under Muhammad's authority. The Prophet Muhammad distributed the booty among his followers, and he kept a large portion of it for himself (Koran 59:1–17).

In the same year of 628, Muhammad directed to a limited number of his followers to get an animal for sacrifice for a pilgrimage to Mecca. In the month of Shawwal, the pilgrimage (Arabic: Umrah) headed to Mecca. The Quraysh, after hearing of the advancing pilgrimage to Mecca, sent out cavalry to stop them. Muhammad changed the route to Mecca, and the pilgrimage reached al-Hudaybiyya outside of Mecca. The negotiations started, and both sides agreed. The agreement is called "Pledge under the Tree" or "Pledge of Acceptance." This agreement included the following items:

1. The contract should last ten years.
2. Deferral of Muhammad's pilgrimage to the next year.
3. Termination of hostilities.
4. Sending back any Meccans that immigrated to Medina without permission of their protector.

The majority of the pilgrimage was not happy with this agreement, but Muhammad got a revelation of "Al-Fath" (Koran 48:1–29) and assured the pilgrimage that this agreement was considered by God a victorious one. After this agreement, Muhammad eradicated the rest of the Jewish tribes in Medina.

The agreement of al-Hudaybiyya was enforced for two years. After two years, al-Hudaybiyya's settlement started a war between two Arabian tribes of Banu Khuza'a and Banu Bakr. Banu Khuza'a had a good relationship with Muhammad. Banu Bakr got weapons from Quraysh and raided the Banu Khuza'a, and a few members of the Banu Khuza'a were killed. As a result of this event, Muhammad sent a message to Mecca and asked the following:

1. Disavow themselves from the Banu Bakr.
2. Pay the blood money for those who killed the Banu Khuza'a tribe.
3. Declare the agreement of al-Hudaybiyya null.

The Meccans accepted the last item of the message. Muhammad began to prepare for the raid to Mecca in 630 CE and assembled a ten-thousand-strong army. Muhammad marched to Mecca and seized it without significant casualty.

At this time, while the Muslims were on the rise, the Persian Empire was weakening due to internal conflict in the dynasty, and the Byzantine Empire was weakening too. The condition was ripe for Muslims to create a new empire.

On June 8, 632 CE, Muhammad, at the age of sixty-two, was poisoned by his wife Safiyya, the daughter of Huyayy ibn Akhtab. Safiyya's second husband, Kinana, was the chief of the Jews in Khyber and was savagely tortured and beheaded under the supervision of Muhammad. Muhammad spent the rest of his illness with Aishah and was buried at al-Masjid an-Nabawi, or the Mosque of the Prophet. This mosque was the first structure Muhammad built in Medina.

Abu Bakr: Abu Bakr (Abu Bakr al-Siddiq, 573–634 CE) became the first caliph, rightly guided (Arabic: Khalifah or Rashidun), or first ruler of the Islamic state. Abu Bakr was Muhammad's advisor, his first convert to Islam, a companion on the journey to Medina in 622 CE, and his father-in-law. Abu Bakr was the closest friend to Muhammad, and offered Aisha, his youngest child, at the age of nine. Aisha had to marry Prophet Muhammad in a small marriage ceremony. Also, after victory in the Battle of the Trench, Muhammad made two more expeditions to the north successfully.

While returning in these years, an accusation of adultery was made against Muhammad's wife Aisha, daughter of Abu Bakr.

The story of Aisha's adultery can be traced in (Koran 24:1-64). Aisha left the howdah to find her missing necklace, and her slaves mounted the howdah for her scheduled travel. The caravan departed without Aisha's presence. She remained in the camp until the next day, when Safwan ibn al-Muttal al-Sulami, a companion (Arabic: sahaba) and a constituent of Muhammad's army found Aisha and brought her back to Muhammad.

Rumors spread very fast in the newly established Islamic community that Safwan and Aisha had committed adultery, particularly by Hassan ibn Thabit, Hammanah bint Jahsh, and many more critical individuals in the city. Some supported Aisha and blamed Ali Ibn Abu Talib, who flatly announced there are plenty of women, and one can change one for another. Muhammad found himself in a challenging position. He was a fantastic politician, and he issued a declaration that he received a revelation of Aisha's innocence and the accusers were subjected to retributions of eighty lashes (Koran 24:1–64). The story of Aisha's adultery shows the position Abu Bakr established in the community; Muhammad could not hurt him by punishing Aisha.

On June 8, 632, the day of Muhammad's death, there was a severe crisis in Medina and elsewhere for the selection of the next leader. In Middle Eastern culture, the heir of the throne is only the son in succession. On this day of intense fight for power, some supported the leadership of Ali Ibn Abu Talib, Muhammad's cousin and son-in-law, because Muhammad treated him as his son; some wanted Abu Bakr, and some followed other leaders. Abu Bakr's political backers were stronger, and they selected him as the first caliph or ruler of the Islamic state. The political backers of Ali Ibn Abu Talib did not vote for Abu Bakr, and some others did not recognize him as a legitimate leader. The political backers of Ali Ibn Abu Talib created the first denomination in Islam and called it Shia. The political backers of Abu Bakr were called Sunni.

Abu Bakr was able to defeat the tribal, political, and religious rebellions known as Wars of Apostasy or the Wars of Riddha and brought them under Muslim control. In reality, after Muhammad's death, several leaders proclaimed God elected them as the prophet and claimed receiving revelations from God through the archangel Gabriel. Being a prophet and receiving revelation became part of tribal cultures. Abu Bakr had no choice other than to eliminate them. He was successful with those campaigns.

He also initiated the first compilation of the Koran after the deaths of several Koranic reciters in the Battle of Yamama. There were concerns that parts of the text were lost, or the possibility that they could be lost. The reign of Abu Bakr was short; he died on August 23, 632 CE, in Medina. Some attribute his cause of death to poison and some to natural causes. Just before his death, he advised the Muslims to accept Umar as a successor, and Muslims did so without incident.

Umar ibn al-Khattab: Umar ibn al-Khattab (or in short, Umar) was born in 586 CE in Mecca and died November 3, 644 CE, in Medina. Umar's position in the state was determined as second in line based on the following reasons, despite some who supported Ali ibn Abu Talib as first in line.

1. Umar gave his daughter Hafsah as a gift to Muhammad in 625 CE to keep himself close to Muhammad.
2. The relationship between Ali ibn Abu Talib and Abu Bakr was extremely sour. It can be seen as the story of Aisha's adultery surfaced in the community; Ali Ibn Abu Talib flatly announced there are plenty of women, and one can change one for another.
3. Abu Bakr, as the head of the state, nominated Umar as his successor before his death.
4. Ali ibn Abu Talib agreed with the selection of Umar, and he gave one of his daughters, Umm Kulthum bint Ali, to Umar in marriage, but he still did not trust his son-in-law.

On August 23, 634 CE, Umar became the second caliph, rightly guided (Arabic: Khalifah or rashidun), or the second ruler of the Islamic State. Umar was literate, and also, once the group of immigrants settled in Medina, he learned politics, government, and military techniques. He was the right hand to Muhammad, and during the ghazu or bounty raids, looting of caravans and annihilating the Jewish tribes, he added to his skills and experience. During his reign, he created a policy for administering the invaded lands. Later in his reign, he established the diwan to provide warriors' pensions to modify the barbarian culture of ghazu or bounty raid and looting as a future invader of the Islamic empire. He established the battalion or garrison of Basra and Kufa in Iraq and Al-Fustat in Egypt. He created the Islamic Hijri calendar and the office of Judge of Gadi.

Umar executed the expansion of Islam's authority by territorial acquisition or by the creation of economic and political dominance over other countries. He used savage force with the Arabian culture of the ghazu or bounty raid. Umar's military had one principal goal: to loot the invaded country and come back home with lots of booty, including women for sale. The Islamic army invaded Syria, Iraq, Egypt, and, most importantly, the Persian Empire was Umar's primary goal.

The Persian Empire was vast, wealthy, and with an abundance of resources. During the rise of Islam, both the Persian Empire and Byzantine Empire were vulnerable because of a long and devastating war between the two empires from 602 CE to 628 CE. Besides, the Persian Empire faced internal turmoil from 628 CE to 632 CE. During this period, Khosrow II, the Persian emperor, was killed by his son Sheroe (or Kavad II) in 628 CE. This turmoil continued to 632 CE, until Khosrow's eight-year-old grandson, Yazdegerd II, ascended to the throne. Still, the empire was very weak, and the military was not led by competent generals.

Umar was persistently attacking the Persian Empire from 633 CE until the complete invasion of the empire in 651 CE. Umar's military savagely killed thousands of people, including children, and imposed heavy taxes (jizya) on those who did not convert to Islam. At the order of Umar ibn al-Khattab, his commander destroyed the treasure of the world, the Academy of Gundishapur, and its library. The piles of books and other written materials were burned. A few that escaped this veracity were later translated into Arabic and modern Persian. Also, all the cities, including Ctephone (the capital city), bridges, palaces, and many magnificent imperial Persian gardens, were burned to ashes.

Under Umar's rule, the Arabs invaded Egypt, Palestine, Syria, Iraq, and especially the most prominent victory was the defeat of the Persian Empire. Umar did not have a divine mandate to conquer other countries. Umar and his military acted exactly as other invaders, like Genghis Khan, who slaughtered hundreds of thousands of innocents at sword-point. Umar continued with all the atrocity of killing a hundred thousand innocent men, women, and children, looting, imposing high taxes, and enslaving men, women, and children. Persians were severely hurt, looking for revenge.

On October 31, 644 CE, Piruz Nahavandi (also spelled Pirouz Nahawandi) (Arabic: Abu Luluha) was a Persian Sasanian solder captured and enslaved by Arabs in the Battle of al-Qadisiyyah. Piruz attacked Umar for revenge during morning prayer, stabbing him many times, and made him bleed profusely. While Piruz was trying to flee, he wounded twelve more Arabs, some of whom later died. Piruz slashed himself with his sword to commit suicide, dying with honor rather than at the hands of the barbarian Arabs. Umar died three days later on Wednesday, November 3, 644 CE. He was buried next to Al-Masjid al-Nabawi, alongside Muhammad and Abu Bakr.

Uthman ibn Affan: Uthman ibn Affan was born around 579 CE in Mecca to the Quraysh tribe (Banu Umayya) and died in Medina on June 17, 656 CE. Uthman was elected as the third caliph, rightly guided (Arabic: Khalifah or rashidun), or the third ruler of the Islamic state, by a council of six men, including Uthman and Ali ibn Abu Talib, who were selected by Umar before his death. He was chosen because he promised to continue the policies that Abu Bakr and Umar were following. During the first year of his rule, he did not show energy and initiative, and he was somewhat a weaker character than his predecessors, but the Muslims conquered new territories. In North Africa, they reached Libya; in the east, they reached as far as the Oxus River and Sind River into India.

Umar tried to make the Koran a central sacred text before his death, which was controversial at the time. According to tradition, Umar commanded Zaid ibn Thabit, who was one of Muhammad's recorders, to collect the various parts together. Umar was killed, so he did not succeed.

Uthman again started this controversial action. At this time, in every Muslim center, like Damascus, Basra, Kufa, Al-Fustat, and many other governorships, Koran reciters were using their version, and they were enraged.

At the time when Uthman came into power, in all the Islamic centers, like Damascus, Basra, Kufa, Al-Fustat, and many other governorships, Koran reciters were using their own version.

Uthman insisted that only one version of the Koran should be used all over the empire. All the governorships became enraged. He was scared that the emergence of the Islamic Empire would fall apart if it did not have a sacred text recognized all over the empire. Uthman personally headed a large committee that established the final version of the Koran. At the end of his reign, the final version of the Koran was completed and copied, and a large number of copies of the Koran were sent to every city and garrison towns. He directed the authorities that the variant versions of the Koran should be burned and destroyed.

Uthman tried to create a stable centralized government to remove the loose tribal alliances of the past. He assigned the critical function of government, like governors and generals, to members of his family. Much of the treasure from taxes and jizya received by the central government was paid to the Uthman family, governors, and generals.

He also tried to reduce the amount of plunder and booty for the military, as well as massive reduction from the central government's treasury. The disgruntled military from Iraq and Egypt traveled to Medina to address their anger and discontent and claim their dues directly with Uthman. He listened and promised to pay their dues. When the military left, he changed his mind on the promise he made. In 656 CE, the military returned to Medina and besieged Uthman's house for more than twenty days. Finally, some of the military broke into Uthman's house and killed him, and he is buried in Medina. The rebellion military acclaimed Ali ibn Abu Talib as the new caliph.

Ali ibn Abu Talib: Ali ibn Abu Talib (or Ali ibn Abi Talib, or Ali, for short) was born around 600 CE in Mecca, Saudi Arabia, and died or was assassinated in 661 CE in Kufah, Iraq. Ali was the cousin and son-in-law of the Prophet Muhammad, and fourth caliph, rightly guided (Arabic: Khalifah or rashidun), or the fourth ruler of the Islamic state by the rebellion military and supporters of Ali. Ali was the first imam of Shiite denomination and the fourth caliph of the Sunnis.

Muhammad, Ali's father-in-law, knew that he was poisoned and would not survive; he did not recommend Ali as his successor. He died in Aisha's care, who was firmly against Ali. It can be seen as the story of Aisha's adultery surfaced in the community, Ali flatly announced there were plenty of women and one can change one for another, and Muhammad ignored him. Ali believed Prophet Muhammad intended Ali to succeed him, more so than Abu Bakr. It was the reason for the split into the Shiite and Sunni denominations.

During the reign of Ali, from 656 CE until his assassination in 661 CE, it was the stormiest period of his life. Ali was brought into power by his supporters and a rebellion military coup. He did not prosecute the murderers of Uthman because they brought him into power. He purged Uthman's supporters from the government. Many members of the Quraysh renounced him because he sided with the Hashemites, a clan of Quraysh to which he and Muhammad belonged.

The selection of Ali by the rebellion military coup was a clear message to all appointees of Uthman that they would be purged soon. Among the appointees were Muawiyah, a close relative of Uthman and governor of Syria. Muawiyah declared the right to take revenge for Uthman's death. Ali and his supporters assembled an army in Kufah, which became his power base and capital.

In 656 CE, Ali faced the Battle of the Camel against an army headed by two senior companions of Muhammad, Talhah and Zubayr, and Muhammad's wife, Aishah. In this battle, Ali killed both Talhah and Zubayr, and Aishah safely escaped back to Medina. After the success in this battle, Ali paid attention to Muawiyah.

In 657 CE, Ali engaged in the Battle of Siffin against Muawiyah. Muawiyah's forces were on the verge of defeat, but he was a smart politician; he followed the advice of his supporter Amr ibn al-As. Muawiyah ordered his military to attach the pages of the Koran on their lances and asked Ali to resolve the conflict by the citation of the Koranic rules. Ali agreed to arbitration.

Muawiyah selected his best supporter, Amr ibn al-As, and Ali picked his best supporter, Abu Musa al-Ashari. The mediation was not in favor of Ali, and the group of arbitrators advised that the arbitration was against the teaching of the Koran and rebelled against Ali. The group of arbitrators did not declare Ali the rightful caliph because the rebellion military coup installed him.

The group of arbitrators left Basra and Kufah and assembled an army at Al-Narhawan. In 658 CE, Ali's army defeated the group of arbitrators' army. Ali's supporters called it the Seceders or Kharijites (Arabic: Khawarij).

A group of Kharijites tried to assassinate those who struggled for power, Muawiyah and Ali. Muawiyah escaped, and Ali was killed on January 26, 661 CE, with a poisoned sword. He died two days later and was buried in Al-Najaf. There was no divine mandate from Abu Bakr to Ali that he should be the caliph or to conquer other countries. It was just struggling for survival and power. Each caliph, including Prophet Muhammad, were killed for their unfairness, atrocities, genocides, and looting by those who suffered and lost loved ones.

Muawiyah ibn Abi Sufyan: Muawiyah I for short was born in 602 CE in Mecca, and died on April or May 680 CE in Damascus. After Ali's assassination in 661 CE, Muawiyah I began his rule as the first Umayyad caliph. Muawiyah I reinstated the unity of the Muslim empire, and he moved the capital of the empire from Medina to Damascus. Muawiyah I died of natural causes in 680 CE; he appointed his son Yazid I as the second caliph of the Umayyad dynasty. The Umayyads lasted until 750 CE in Damascus, and were ousted by the Abbasid dynasty. The Abbasid dynasty governed from 750 CE until the rise of the Mongols under cult leader Genghis Khan, who started just like Prophet Muhammad with killing hundreds of thousands of innocent men, women, and children, and looting. The Abbasid empire was destroyed by the Mongol invasion in 1258.

Jalal al-Din Rumi: Also called by the honorific Mawlana Jalal al-Din Rumi, or Mawlana Jalal al-Din Mohammad-e Balkhi (in short, Rumi or Balkhi), Rumi was born on September 30, 1207, in Balkh, Afghanistan, and died on December 17, 1273, in Konya, Turkey. He is known in the west as Jalal al-Din Rumi because his family settled in the Anatolia (Eastern Roman Empire); therefore, he got the surname Rumi. He was known in the East as Mawlana Jalal al-Din Mohammad-e Balkhi. In this section of the book, Jalal al-Din Rumi will be called Rumi.

Rumi's father, Baha al-Din-e Valad, also known as Baha Walad, was a mystical theologian, preacher, religious scholar, a Sufi teacher, and an expert in Hanafi law.

Rumi was born during a turbulent time. Because of either the rise of Genghis Khan in the east and Crusaders in the west, or disagreements with a local ruler, Rumi's entire family, including Baha al-Din-e Valad, and his disciples, fled Balkh between 1214 and 1220 to the west.

First, they were stationed in Nishapur, where the family met Farid al-Din Attar, a prominent Persian poet who consecrated young Rumi. After a long journey in the Middle East, the entourage reached the Anatolia. They stayed for a while in Laranda (now Karaman), where Rumi married Gowhar Khatun in 1225. His first son was born. He had two sons from Gowhar Khatun, Sultan Walad and Ala al-Din Muhammad.

After the death of Gowhar Khatun, he remarried and had a son, Amir Alim Chalabi, and a daughter, Malekeh Khatun. In 1228, Baha al-Din-e Valad was called by the sultan of the Seljuks to the capital in Konya to teach theology to numerous religious schools. In 1231, Baha al-Din died, and he was succeeded by his son Rumi.

Rumi was born, raised, educated in a very religious family of Sufi mystics and the Hanafi sect of Islam. He was a Sufi mystic whose spiritual goal was to have a direct, personal experience with God. After the death of his father, Rumi filled the vacuum and was teaching theology in several religious schools, lecturing to his disciples. Also, he was a unique poet in the Persian language, very close to the level of Shakespeare and American poets Walt Whitman and Ralph Waldo Emerson. He was prominent for the instructive epic of his lyrics in *Masnavi (Mathnawi) Manavi (Spiritual Couplets)* in *six volumes*, composed in Konya and regarded as one of the greatest poems of the Persian language; it was called Koran in Persian. This work influenced mystical thoughts and literature around the world, mainly in Islamic countries. Rumi was multilingual; he used Persian, Arabic, some Turkish, and a little Greek in his poetry.

A sudden change in Rumi's life marked him for greatness on November 30, 1244. Rumi met the wandering itinerant religious and mystic man named Shams al-Din Muhammad (or Shams al-DinTabrizi) in the street of Konya. Rumi and Shams from that moment on became inseparable. For months, the two mystical friends lived together. Rumi did not attend his teaching schools; he neglected family and disciples. In turn, his family and disciples forced Shams to leave Konya in February 1246. Rumi was devasted and heartbroken with the absence of Shams. His eldest son, Sultan Walad, searched for Shams and brought him back to Konya from Syria. Rumi came back to life at seeing his beloved. His family and disciples could not tolerate the existence of Shams and his close relationship to Rumi.

In 1247, Shams disappeared from Rumi's house forever. Later, it was found that Shams was murdered by his sons, who hurriedly buried him close to a well that still exists in Konya. The power and experience of same-sex love, yearning, and loss of his lover changed Rumi into the greatest poet that the Islamic world had ever generated. Not only did he turn into a great poet, but he also brought a revolution in the Islamic Sufi community so that the forbidden music, science, and dance became a routine ritual. His ghazal poems, approximately thirty-five thousand Persian couplets and two thousand Persian quadrants, reflect the homoerotic (revealing or portraying the same sex or homo desire) in different stages of love. From his closeness to Shams, the intensity of their same-sex love is indisputable.

In around 1248, Rumi was grieving the loss of a loving friend in verse and whirling dance. The complete assertion of his homo desire was expressed by his inserting Shams's name instead of his pen name at the end of most lyrical poems.

His burial procession was attended by a vast crowd of many nationalities and faiths to the Green Dome in Konya. The Green Dome is still a place of pilgrimage for Turkish people and tourists.

Rumi's major works are the *Masnavi (Mathnawi) Manavi (Spiritual Couplets) in six volumes*. It contains about twenty-seven thousand lines; *The Divan-e Kabir (Great Work) or Divan-e Shams-e Tabrizi (The Works of Shams-e Tabrizi)*, with about thirty-five thousand Persian couplets and two thousand Persian quadrants. Besides these, there are some lectures and talks, sessions, and letters. Some of his more homoerotic poems were not published until recently. Here is Rumi's poem for Shams:

> When I am with you, we stay up all night.
> When you're not here, I can't go to sleep.
> Praise God for those two insomnias!
> And the difference between them.

Scriptures

The *Koran* (English: recitation) is the central scripture of Islam. Muslims believe the Koran was revealed through God's messenger Muhammad in 610 CE by the archangel Gabriel (Arabic: Jibril). The revelations continued until Muhammad's death on June 8, 632 CE. The Koran is organized in 114 surahs (English: chapters), which are written in the old Arabic dialect. Each surah contains a varying number of ayats (English: verses). The Koran is not organized in the chronological order in which the revelations were revealed. It is well-ordered following the short surah.

1. The Al-Fatihah (Prologue) is the longest surah (1).
2. Al-Baqarah (Cow) with 286 verses to the shortest surah.
3. Al-Kauthar (Preeminence) with three verses (108).

The Koran is also divided into 30 equal juz (English: parts). These divisions are easy for Muslims who want to recite the Koran in a couple of hours. The thirty parts are divided between a couple of reciters, and then all the reciters recite at the same time. The Koran has a total of 6,666 verses by Islamic scholars and a total of 6,236 by others.[116] In 1500, Muslims were not able to determine the number of verses correctly.

Prophet Muhammad was illiterate, in general; God directed revelations orally to Muhammad through angel Gabriel. After Muhammad's death, creating a scripture was not a priority. Abu Bakr was busy trying to bring peace in Mecca and other parts of the Arabian Peninsula. During the reign of Omar, after the establishment of the Islamic empire, Muslims needed to have a scripture. Very little of the Koran was written on materials like stone and palm branches. Making the Koran a central sacred text was a controversial subject, but Omar assigned Zayed ibn Thabit to collect the Koran. He copied out on sheets of parchment whatever he found and handed them over to Omar. After Omar's death, the written collection was kept by Omar's daughter Hafsa.

Uthman again started this controversial action. At the end of his reign, the final Uthman version of the Koran was completed. Some scholars think that certain surahs were deleted or altered, and especially many critical attacks on the Umayyad tribe may have been removed at Uthman's instruction.

The Islamic sacred scripture, the Koran, is the will of God to which Muslims must surrender.

In reality, scholars believe the Koran is in two parts or books:

1. *The Older Koran or the Meccan Phase*, before the Hijra on July 16, 622 CE. This part contains almost thirteen years of revelation.
2. *The Later Koran or the Medinan Phase*, started from 622 CE to 632 CE. This part includes nearly ten years of revelations.

The Older Koran or the Meccan Phase started in Mecca, and the purpose was to convince and preach to a small audience including Jews, Christians, and Zoroastrians with simple surahs of fear of hell (Arabic: aljahim) and rewards in paradise (Arabic: aljana) (Koran 1:1–7). The surahs are faithful to the concept of the resurrection (Arabic: Qiama) and Judgment Day (Arabic: yawm al-diyn), and Judgment Day is soon to come (Koran70:6–23). Later surah corrects this from near to an unknown, prolonged term (Koran 72:25). The Older Koran is all about spirituality, tranquility, tolerance, and expressing closeness to all religions and respect for them.

As Muhammad arrived in Medina, he continued preaching and convincing the new audience for conversion. He stood by the tranquility, tolerance, and peaceable words of the Meccan Phase. After Muhammad and his followers settled, financial aid and food from hosting communities reduced and stopped. Muhammad and his followers started to raid and loot the trade caravans, and he praised their followers for submitting to his demand. The result of raiding and looting of the caravans and converting the hosting communities was the beginning of conflicts between Jews and Muhammad. During the escalation of this conflict, Muhammad revealed the Later Koran or the Medinan Phase.

The following changes can be seen:

1. Early revelations describe Muhammed as an admonisher whom God has sent revelations to admonish his close audiences (Koran 32:3). In another revelation, it shows him as only God's messenger (Koran 11:57).
2. Medinan revelation directs his followers to obey Muhammad and his judgment (Koran 4:59–70), and in one revelation, God proclaims Muhammad as quintessence for the believer (Koran 33:21). This revelation gave Muhammad the power to direct the followers going to war in Medina against opponents.
3. The Meccan revelation about Jews was limited to only ancient Israelites' sin. Muhammad was welcomed by Jews in Medina. Jews were providing Muslims food and other means of living in Medina. Soon, Muslims in Medina became stronger and needed land, property, and business. Medinan revelations were a direct attack and critique of Jewish belief. The following revelation came to Muhammad: a divine authorization to kill the Jews (Koran 33:26), showing animosity with Jews (Koran 5:64–65), and more (Koran 5:19, 4:154–155, 9:30–31).
4. Medinan revelations have detailed legal regulations. The story of Aisha's adultery can be traced in Koran 24:1–64. (See Abu Bakr in "Leaders" for details.)

5. Comparing both parts of the Koran.

The Older Koran or the Meccan Phase says, "Bear with patience what they say, and gracefully come away from them" (Koran 73:10).

The Later Koran or the Medinan Phase says, "Kill them wherever you overtake them and expel them from wherever they have expelled you" (Koran 2:191).

The Older Koran or the Meccan Phase says, "O Lord, these are people who do not believe. Turn away from them and say, 'Peace.' In the end, they shall know their foolishness" (Koran 43:88–89).

The Later Koran or the Medinan Phase says, "When you meet the unbelievers, strike off their necks until, when you have inflicted slaughter upon them, carefully tie up the remaining of captives, and ransom them heavily" (Koran 47:4).

The Older Koran or the Meccan Phase says, "Tell the believers to forgive those who do not fear the visitation of Allah, so that He may avenge the people for their deeds" (Koran 45:14).

The Later Koran or the Medinan Phase says, "Against them make ready your strength and steed of war to the utmost of your power, to strike terror and fear in the hearts of the enemies of Allah and your enemies. Whatever you send in the way of God, He will pay you back" (Koran 8:6).

The Older Koran revelations were all about reconciliation, peacemaking, tranquility, spirituality, and did not express the authority of God. The Later Koran revelations were all about occupation, killing, and showing the authority of God and Muhammad.

The Koran's contents are God's messages revealed to Muhammad. God doesn't make mistakes or change his mind. Only humans can make mistakes and change their minds. In the Koran, there are hundreds of contradictions; a clear example of these contradictions are found in 9:17, 9:69, 99:7, 2:62, 9:28-33, 5:17, 5:72-73.[117]

Hadith or Hadit (English: story and narratives) consists of tradition or sayings of Muhammad and his companions, which is a significant source of moral guidance and religious law. Hadith is the scripture to the authority of the Koran. During the three hundred years after Muhammad's death, six significant collections were compiled by different individuals. Most of them are not considered to be authentic. Imam al-Bukhari collected the final version in 810–870 CE. Based on the perceived validity of scholars' compilations, there are two versions of hadith, one acceptable to the Sunni denomination and the other to the Shia denomination.

Sunna or Sunnah (English: habitual practice, path, or way) is referring to the action of Muhammad, who set the example for all Muslims to follow. The pre-Islamic Arabs' sunnah was referred to examples established by the tribal ancestors and accepted as normal to entire Arab communities. After Muhammad died, Muslims did not accept the sunnah. They looked first to the people in Medina, Hejaz, and other provinces. Finally, after reconciling the differing community practices, Abu Abd Allah al-Shafii approved the sunnah as a legal status second to the Koran.

Denominations

After Muhammad's death, Islam split into two main denominations: Sunni and Shia. Sunnis wanted succession based on political terms and rank, while Shia wanted a sequence based on spiritual terms. Later, other subdivisions took place within the Sunni and Shia, like Druze, Alevi, Alawis, Ahmadiyyas, Nation of Islam, and Ibadi, which will not be discussed in this book.

Sunni is the largest of the two main branches of Islam. Sunnis understand the Sunnan (English: tradition, way, or pain), which is the way of Muhammad. The Sunnis recognize the first four caliphs as the rightful successors of Muhammad.

Sunnis are about 85 percent of the Muslim population, which equates to a population of 1.5 billion people.

The Sunni denomination is divided into the following schools: Shafii, Hanbali (includes Wahabi and Salafist), Maliki, Hanafi (includes Barelvi), and Deobandi. T'aiiban comes from Deobandi.

Shia are followers of Ali or party of Ali (also called Shiite). As a group, Shia is a smaller branch of Islam. Shia Muslims believe that

Muhammad intended his son-in-law and cousin, Ali ibn Abu Talib, to become his successor. However, the friction between Muhammad's followers existed during the life of Muhammad. Between Ali and Aisha, Muhammad's wife and Abo Bakr's daughter survived a huge rift. Muhammad died in the presence of Aisha and Abu Bakr. Therefore, he became the fourth successor of Muhammad with the military coup. Most Shias view Ali as the second prominent figure in Islam.

The Shia denomination is divided in the following schools: Twelvers, Ismaili, Alevi-Bektashim, and Zaidi. The Twelvers or Twelve imam divinely set spiritual leaders are the following:

1. Ali ibn Abu Talib
2. Hassan ibn Ali
3. Hussain ibn Ali
4. Zayn al-Abidin
5. Muhammad al-Baqir
6. Jafar al-Sadiq
7. Musa a-Kazim
8. Ali al-Rida
9. Muhammad al-Jawad
10. Ali al-Hadi
11. Hassan al-Askari
12. Muhammad al-Muntazar or last imam

The Twelvers are the descendants of Ali ibn Abu Talib, and the last descendant, Muhammad al-Muntazar al-Mahdi, was the twelfth imam. Shia believes that al-Mahdi was hidden in a cave in Samarra to stay away from persecution. According to the Shia belief, he has been hidden by God until his return to the final judgment or the end of time.

Kharijite (Arabic: Khawaij) comes from the Arabic word *kharij*, which means to leave or get out. Kharijite is the first Islamic denomination beginning with a dispute over the caliphate. After the death of Uthman, the rebels selected Ali ibn Abu Talib as the fourth caliph. In the indecisive Battle of Siffin against Muawiya's forces, Ali agreed to arbitration. The Kharijite opposed the arbitration, and they believed that the judgment belongs only to God. Later, Ali campaigned against Kharijite and killed many of them until Kharijite assassinated Ali. Kharijite thought that anybody, even a Black slave that met the requirement, could be elected as caliph. Besides, they were known for being fanatics and puritans.

Sufism is a mystical Islamic group of people who try to experience God directly, especially by meditation and prayer. Islamic mysticism (Arabic: tasawwuf; English: to dress in wool) has been called Sufism in Western languages since the nineteenth century. Sufism started for the first time as a reaction to the worldliness of the Umayyad period (661–749 CE). Sufism continued developing until the rise of fraternal order under Jaiai ai-Din Rumi (Maulana Jalal al-Din Mohammad-e Balkhi). He was a revolutionary Sufi and had no belief in divinity, but he could express his real feeling for fear of death. (See Jalal al-Din Rumi in "Leaders.")

Beliefs

Muslims believe in the oneness of God who assigned Muhammad as the last prophet to convey the divine message to his people. All Muslims from every denomination share the following beliefs.

Deity: God (Arabic: Allah), the only one God, the creator, sustainer, and the restorer of the world in Islam. Allah is not a different God from the Judeo-Christian God. Pre-Islamic Julian and Christian communities on the Arabian Peninsula called their God Allah. Allah is the central point of the Islamic faith. Muslims believe Allah is unique (Arabic: Wahid) and natural (Arabic: Ahad). Allah is the creator of the universe, rewarder, forgiver, judge, and is omnipotent, omnipresence, omniscient, and omnibenevolent.

Prophets: Islam teaches that God elected human messengers or prophets to deliver his messages to their people in different times and places, since the beginning of time. In the Islamic religion, Muhammad is the last in succession of 124,000 prophets. Twenty-five prophets are named in the Koran, including Adam, Noah (Arabic: Nuh), Abraham (Arabic: Ibrahim), Moses (Arabic: Musa), Solomon (Arabic:

Sulayman), and Jesus (Arabic: Yasue or Messiah) (Koran 33:40). The 124,000 prophets were sent by God for the guidance of the Middle East only, and the names of those prophets are from Abrahamic religions.

Scriptures: The sacred scripture or the revered text of Islam is the Koran (Arabic: recitation). About the format of revelation, there are a couple traditions and myths. In general, God directed revelations orally to Muhammad through the angel Gabriel (Arabic: Jibril). The process of revelation started when Muhammad was forty, and continued until Muhammad died at the age of sixty-two, after being poisoned by his wife Safiyya, whose husband Kinana, as the chief of the Jews in Khyber, was savagely tortured and beheaded under the supervision of Muhammad. Otherwise, the process of revelation could continue until the natural death of Muhammad. The second scripture to Koran is Hadith, the collected sayings of Muhammad, and constitutes another source of guidance for Muslims in addition to the Koran. Hadith was collected several centuries after Muhammad died.

The Five Pillars of Islam: The Five Pillars of Islam (Arabic: Arkan al-Islam) are designed to anchor Muslims for the rest of their life. They are similar to the Ten Commitments in Judaism and Christianity. The five obligatory duties on every Muslim are the following:

1. *The profession of faith (shahada)*: There is no God but God, and Muhammad is his prophet.
2. *Prayer (Salat)*: Devout Muslims must pray five times a day toward Mecca, and they are before sunrise, at noon, in the afternoon, immediately before sunset, and after dark at night. This ritual was borrowed from Zoroastrianism.
3. *The Zakat*: Each Muslim must pay an obligatory tax, or zakat, once a year to the state to benefit the needy and poor. Zakat is calculated at 2.5 percent on the categories of property, and they are camel, sheep, cattle, and goats; fruit; food grains; and moveable goods like gold, and is payable after one year of ownership.
4. *Fasting (Sawm)*: Muslims must fast for the month of Ramadan, the ninth lunar month of the year. Fasting begins at daybreak and ends at sunset. During fasting, drinking,

eating, smoking, and sexual intercourse are forbidden (unless sick or a woman during the monthly period).
5. *Hajj (Pilgrimage to Mecca)*: A Muslim who is able financially and physically must make a pilgrimage to Mecca at least once in his lifetime.

Six major faiths or beliefs: The following six articles are those that Muslims must accept as part of faith, or iman, as they are laid out in the Koran and Hadith.

1. *Believe in the oneness of Allah or God*: Muslims must believe that God, the only one God, is the creator, sustainer, and the restorer of the world in Islam.
2. *Believe in the angels of God*: Muslims must believe in the existence of angels who obey God and, like Gabriel, (Arabic: Jibril) brought God's revelation to Muhammad. Also, Muslims must believe in the presence of Satan (Arabic: Iblis), who deceived Adam and encouraged him to eat the forbidden fruit of heaven. Adam was expelled from heaven.
3. *Believe in the books of God*: Muslims must believe that God revealed books to other prophets of the Middle East as well as the last prophet of the Middle East, Muhammad.
4. *Believe in the prophets and messengers of God*: Muslims must believe that God elected 124,000 prophets, 25 prophets are named in the Koran, and the last prophet, Muhammad, was only in the Middle East to guide humanity in the Middle East.
5. *Believe in the Day of Judgment and resurrection*: Muslims must believe that on the Day of Judgment, every Muslim will be judged based on what that person did in life. Those who were good and followed God's guidance go to heaven, and those who were not following God's guidance go to hell. Jews and Islam borrowed the Day of Judgment, with slight modification, from Zoroastrianism.
6. *Believe in God's foreknowledge and his will*: Muslims must believe that life is preordained. (This concept conflicts with "free will" [Koran 72:26–28].)

Creation: Creation in Islam is identical to Judaism and Christianity, with less detail. The Koran states, "We created the heavens (sky or cover) and the earth and all between heavens and earth in six days, and no weariness touched Us" (Koran 50:38). This statement indicates God created the earth and heaven, not the universe. God created the first humans, Adam and Eve, from clay, charming and handsome (Koran 32:7). Before placing them in heaven, Allah directed angels to bow down to humans. Allah led Adam not to eat the forbidden fruit, but in contrary an angel, Iblis, enticed him to do so. Allah was enraged at Adam for disobeying him and cast the first human and Iblis from heaven and sent them to earth. Iblis is an opposing force to Allah and will misdirect humans forever.

The concept of human creation from the clay was known to Arabs from the Mesopotamian religion, and especially Zoroastrianism. The Zoroastrian god Ahura Mazda (the Wise God) created the first man from clay, tall and handsome, and gave life to him. The Iblis or evil is identical to the opposing power of Angra Mainyu to the Ahura Mazda (the Wise God) in Zoroastrianism.

Iblis: According to the Koran, after the creation of Adam, God ordered all the angels to bow down to Adam in homage. They all bowed, but Iblis was not among those who bowed, and conversation and quarrel between God and Iblis continued. Iblis or Shaitan (English: devil) is named Satan in Judaism and Christianity. He was expelled from heaven, and God gave him the authority to act as an enemy against humanity and live to the Day of Judgment (Koran 7:11–17). God says Iblis is an enemy to you, so hold him as a foe (Koran 35:6).

The above narrative is remarkably close to Zoroastrian belief about the creation of the first man by Ahura Mazda (the Wise Lord) from a clay called Gayōmart, tall and handsome. Ahura Mazda called on Angra Mainyu (the Evil Spirit) to help the creature and praise him. Conversation and argument between Ahura Mazda and Angra Mainyu continued as did between God and Iblis.

Islam borrowed not only the idea of Iblis, but also many more ideas from the Zoroastrian religion and modified them to meet their political and religious goal to include them in the Koran.

The Soul: God created the first humans from clay and then animated them by blowing his soul into them. The soul that entered the body of humanity was pure (Koran 15:29, 38:72). God takes humans' eternal soul at night and gives it back to them when they wake up (Koran 6:60–61). The immortal soul will be taken from the body by the angel of death, or Azrael (Arabic: Malak al-Maut), and give it to God, waiting for final judgment. The soul, while waiting for final judgment, stays in Barzakh (Persian: Barzakh means barrier or partition) and experiences heaven or hell based on performance in a lifetime. Muslims very likely borrowed the soul from Zoroastrianism, with minor modification.

Afterlife: Muslims believe in the final judgment or the end of time when all life is annihilated, and is followed by the resurrection of all deceased and jinn or devils and judgment by God, according to their deeds (Koran 17:100). During this time, the hadith describes that chaos and corruption rule the world by the Al-Masih ad-Dajjal pretending to be the messiah before the end of time. The messiah comes to the earth, defeating the Al-Masih ad-Dajjal and installing a government of peace, and frees the world from cruelty. On the Yawm ad-Din, every human must pass a narrow bridge (Arabic: as-sirat). Those found worthy based on their deeds cross the narrow bridge and enter heaven (Arabic: Jannah), and those found unworthy fall from the bridge into hell. At the center of hell (Arabic: Jahannam) is the tree of Zaqum; the fruit of the Zaqqum tree boils in the stomach, causing extreme pain. Islam borrowed the narrow bridge mentioned in the Koran from Zoroastrianism.

Rituals and Customs

Religious Islam, like other religions, has many rituals and customs. The countries outside of the Arabian Peninsula before the invasion of Islam had their rituals and customs, and those countries have a mixture of Islam and their rituals and traditions. The prevalent ritual is the Five Pillars.

Imam: Imams (English: leader) are scholars who lead the Muslim community in prayers, Jummah (English: Friday) service, sermons, and teach the Koran. The origin of the office of imam originated differently in different denominations. Sunnis are considering imams as a continuation of khalifah (English: caliph) or the successor of Muhammad, who assumed his political and administrative, but not religious, functions. Shiites considered Ali as the successor of Muhammad. Ali tried to preserve the leadership of the entire Muslim community, along with the decedents of Muhammad. Imams in the Shiite denomination possess the same expertise and authority as Muhammad's direct lineage.

Mosque: A mosque (Arabic: masjid or Jami) is a house of prayer. The meaning of the Arabic word *masjid* is a place of prostration to divine or God. The mosque is used for prayers like Friday prayer, worship, prayer for deceased Muslims, and Eid prayer. There are two types of mosques. The Masjid Jami or collective mosque is controlled by the state, and the other one is a smaller masjid, operated privately by various people within the community. Every mosque has a prayer niche in the qiblah wall facing the direction of Mecca and is usually ornately decorated and are in different sizes.

Kaaba: Kaaba or Kabah means cube (also called al-Kabah al-Musharrafah). This is a small cubic shrine located at the center of the Great Mosque in the Hejaz city of Mecca. In ancient times, Kaaba was called al-Qdis (English: holy) and al-Nadhir (English: one who warns). Muslims in any part of the world must pray toward Kaaba and, during the pilgrimage or haji, must visit the Kaaba. Kaaba is about fifty feet high, forty feet long, and thirty-five feet wide. Inside the Kaaba are only three pillars supporting the roof and a few suspended gold and silver lamps. In the eastern corner of the Kaaba wall is the Black Stone (Arabic: al-Hajar al-Aswad), which is not found anywhere in the Koran. This small cube building contains the Black Stone and some idols and was a very important site of pilgrimage for the Arabian Peninsula. Pre-Islamic pagans walked around it seven times.

According to the Koran, God asked Ibrahim to build the Kaaba, and Ibrahim, with his concubine Hagar and their son Ismail, built the Kaaba (Koran 2:125–127, 22:26–27). According to the Torah, Ibrahim was never in Mecca, and he was the first Hebrew who had a covenant with God. As Ibrahim journeyed from Ur to Haran, and from Haran, at the age of seventy-five, he settled at Shechem, the promised land of the Canaanites (Gen. 11:27, 25:10). One of the sacred scriptures must be wrong, the Koran or the Torah. It appears that the Koran borrowed Ibrahim's myth with some revision to meet Muslim legend.

According to the Islamic legend, the black stone was given to Adam by God on his fall from paradise. Some other traditions believe this stone was found by Ibrahim while building Kaaba. This stone is dark black, polished smooth by the hands of countless pilgrims. The original size of the stone is not precise. Many times, it was stolen and burned. At present, the exposed face of it is approximately 8 inches (20 centimeters) by 6.5 inches (16 centimeters) and is five feet (1.53 meters) above the ground. Some pieces of the Black Stone, maybe six, are in Istanbul, Turkey.

Secular historians indicate the history of stone worship; perhaps a meteorite killed someone's beloved child or the head of the family. That stone would then become the object of worship and reverence. Tradition supports that Kaaba was initially built in a simple rectangular shape without a roof to house the Black Stone. It came from space, and therefore, the building was without a roof, in case another stone comes. The latest pre-Islamic reconstruction of Kaaba was performed with alternating courses of wood and masonry by the Quraysh tribe, who ruled Mecca in 608 CE. It indicates that the pre-Islamic Kaaba was the House of Black Stone and not the House of God.

Muhammad, with all the pagans, joined the pilgrimage and walked around it seven times before God selected him as a prophet until his Hijra to Medina in 622 CE. Muhammad chose Kaaba as Muslims' Qibla. The Jews welcomed Muhammad and provided him with housing, food, clothing, a place for prayer, and other means of life. Muhammad was thankful to the Jews and their generosities.

After Muslims built the first mosque in Medina, they changed Qibla from Kaaba to Jerusalem to make the Jews happy. Conflict arose between Muhammad and the Jews; Muhammad altered the direction of Muslims' prayers in 630 CE from Jerusalem to the Kaaba (Koran 2:142–144).

Circumcision: Circumcision (Arabic: Khatna or Khitan) for males is Sunnah (English: habitual practice) in Muslim communities. Circumcision was part of the Arabian culture for males and females. Circumcision wasn't an item of discussion during the life of Muhammad because Muhammad was born circumcised, and for that reason, it is not in the Koran. It is in hadith and sunnah, which was copied from Judaism and was part of Arabian culture. The Islamic male circumcision is similar to Jewish circumcision, but not identical.

Birth ritual: Immediately after birth, the child's father or a family elder holds the child, facing him toward Mecca. He whispers *adhan* (English: to listen) in the right ear of the child, which is a general statement of shared faith and belief. Seven days after birth, the child receives his or her name, and his or her head is shaved. This ceremony is called *aqiqah*, an Arabic tradition. Circumcision is performed between birth and seven years old without significant ceremony.

Marriage rituals: Marriage rituals vary from culture to culture; the same is true between all Islamic denominations. According to sharia, marriage is a legal contract between a man and a woman. Both men and women are to consent to sign the agreement in the presence of two Muslim witnesses. The wedding ceremony is called nikah. During nikah, a discussion takes place between the representative of the bride and bridegroom concerning the amount of gifts (Arabic: mahr). This mahr is payable to the bride in case of divorce (Koran 4:4). After this agreement, depending on the culture, they exchange rings, and the guests receive dinner. Muhammad married nineteen women, but the Koran allows up to four wives (Koran 4:3).

Haram: Haram, also spelled Haraam (English: forbidden or proscribed), means anything prohibited in the Koran or resulting in sin when committed by a Muslim. In the Koran, hadith or sharia is a long list of forbidden items: eating pork and monkey, drinking alcohol, drawing art that resembles a human, gambling, usury or interest on the debt, owning pets other than a cat, music with instruments, dancing, eating or washing with the left hand, premarital relationship, masturbation, homosexuality, talking to people who are homosexual, a woman talking to another man other than her husband, refusing sex when asked by a husband, going outside in public without the husband's consent, arguing with a husband, arguing with the husband's other wife, divorcing a husband, marrying a non-Muslim man, a woman raising her hand against her husband, refusing to cook food, and many more.

Festivals and Holy Days

The Islamic calendar is lunar, like the Jewish calendar and many other religions, like Hinduism and East Asian religions. It consists of twelve months of 29 or 30 days, for a total of 354 days. The last month of the year varies between 29 and 30 days. This method allows the calendar to stay in step with the actual phases of the moon, and therefore, some years have 355 days. Unlike most other lunar calendars, the Islamic calendar is not adjusted to keep in step with the solar year. The first day of the era is July 16, 622 CE, or Muharram 1, 1 AH (anno hijiri or after the Hijra). Muharram is the first month of the year in Islam.

The main Islamic festivals are two, which are set down in Islamic law: Eid ul Fitr and Eid ul Adha. There are some other days that Muslims celebrate.

Al-Hijra (September or October): Al-Hijra, or the New Year, is the first day of the month of Al-Hijra; it is the month in which Muhammad was forced to emigrate from Mecca to Medina on July 16, 622 CE.

Ashura (September): Ashura (also spelled Aashurah, Aashoorah, or Ashurah) is an Islamic holy day that is observed on the tenth of Muharram. Soon after hijra, Muhammad designated a day of fasting from sunrise to sunset to match the Jowish day of Atonement, or Yom Kippur, because the relation of Muslims to Jews was in good standing. Later, the relationship between Muhammad and the Jews went sour, so not only did he change the direction of the prayer from Jerusalem to Mecca, but also, he designated Ramadan the month of fasting. Ashura is the month of mourning and grief for the Shia denomination for the death of Husayn, the son of Ali and grandchild of Muhammad. A lot of Shia lash themselves on the back with sharp steel chains and ritually cut themselves. Shia make pilgrimages to Mashhad al-Husayn in Karbala, Iraq, where the tomb of Husayn was.

Eid Al-Adha (August or September): Eid Al-Adha is the four-day Festival of the Sacrifice, a significant festival that takes place at the end of the Hajj (English: pilgrimage) to Mecca. It is also known as al Eid al-Kabir (English: Major Festival). This festival begins on the tenth day of the last month of the Islamic calendar. This festival celebrates Allah's gift of a ram in place of Ishmael (Arabic: Ismail), whom God had instructed Abraham (Arabic: Ibrahim) to sacrifice. This myth is also in Judaism and Christianity; the child to be sacrificed is Ishmael's brother Isaac (Arabic: Ishaq).

In this festival, the poor do not understand; they don't need to go to Mecca if they can't afford the pilgrimage, but they do anyway. Also, Muslims kill millions of animals; in most cases, they go to waste.

Eid Al-Fitr (May, June, or July): Eid Al-Fitr or Id-Fitr (English: Festival of the Breaking of the Fast) is one of the two significant festivals. This festival starts at the end of Ramadan, or month of fasting.

Laylat al-Qadr: Muslims celebrate the Lavat a-Cadr or the Night of Power on one of the last ten nights of Ramadan, usually the twenty-seventh night. The Night of the Power is when God decided to reveal the Koran to Muhammad through Jibril as a guide to his people. Muslims spend these days and nights, especially the twenty-seventh night, in the mosque, to pray to God for what they wish.

Mawlid (Mawlud, or Milad) (November, December, or January): Mawlid is the birthday of Muhammad (Milad al-Nabi). This day was arbitrarily selected as the twelfth day of the fifth month of Rabi al-Awwal. Most of the Islamic countries have a holiday that involves feasts and stories about Muhammad's life.

Influence of Zoroastrianism on Islamic Religion

See chapter 5, Zoroastrianism, Influence of Zoroastrianism on Judaism, Christianity, and Islam.

Summary

Islam is one of the significant Abrahamic (Arabic: Ibrahim) monotheistic religions of Middle Eastern culture. Muhammad was the founder of Islamic religion on the Arabian Peninsula in the seventh century CE. In Arabic, Islam means submission or surrender.

In the Islamic religion, Muhammad is the last in succession of 124,000 prophets. All these prophets are Middle Eastern. There are no names in the Koran from other races or other parts of the world.

The Islamic sacred scripture, the Koran, is the will of God to which Muslims must surrender. The Koran was revealed through God's messenger, Muhammad, in 610 CE by the archangel Gabriel (Arabic: Jibril). The revelations continued until Muhammad's death.

In reality, scholars believe the Koran is in two parts or books:

1. *The Older Koran or the Meccan Phase,* before the Hijra on July 16, 622 CE. This part contains almost thirteen years of revelation.
2. *The Later Koran or the Medinan Phase,* started from 622 CE to 632 CE. This part includes nearly ten years of revelations.

The Older Koran or the Meccan Phase started in Mecca, and the purpose was to convince and preach to a small audience, including Jews, Christians, and Zoroastrians, with simple surahs of fear of hell (Arabic: aljahim) and rewards in paradise (Arabic: aljana) (Koran 1:1–7). The surahs are faithful to the concept of the resurrection (Arabic Qiama) and Judgment Day (Arabic: yawm al-diyn), and Judgment Day is soon to come (Koran70:6–23). Later surahs correct this from near to an unknown, prolonged term (Koran 72:25). The Older Koran is all about spirituality, tranquility, tolerance, and expressing closeness to all religions and respect for them.

As Muhammad arrived in Medina, he continued preaching and convincing the new audience for conversion. He stood by the tranquility, tolerance, and peaceable words of the Meccan Phase. After Muhammad and his followers settled, financial aid and food from hosting communities reduced and stopped. Muhammad and his followers started to raid and loot the trade caravans, and he praised their followers for submitting to his demands. The result of raiding and looting of the caravans and converting the hosting communities was the beginning of conflicts between Jews and Muhammad. During the escalation of this conflict, Muhammad revealed the Later Koran or the Medinan Phase.

In ancient times, Kaaba was called al-Qdis (English: holy) and al-Nadhir (English: one who warns). Kaaba was a site of pilgrimage for the Arabian Peninsula pre-Islamic pagans to walk around seven times. Muhammad, with all the pagans, joined the pilgrimage and walked around it seven times before God selected him as a prophet. Muhammad chose Kaaba as Muslims' Qibla. After Muslims built the first mosque in Medina, they changed Qibla from Kaaba to Jerusalem to make Jews happy. Conflict arose between Muhammad and Jews; Muhammad altered the direction of Muslims' prayers in 630 CE from Jerusalem to the Kaaba (Koran 2:142–144).

Regardless of how religion is progressive, the primitive faith always has some roots in it; in Islam, it is the black rock in Kaaba and Kaaba itself.

In Islam, there are two primary components of fear and reward after death.

1. *Heaven* is the abode of God, angels, and souls of the righteous after death, and is a land flowing with milk and honey and a life of ease. Heaven is mentioned 124 times in the Koran.
2. *Hell* is a place after death in the afterlife in which the evil person is subjected to punishment and burned and tortured for eternity. Hell is mentioned 228 times in the Koran.

CHAPTER 14

Sikhism

July 8, 2017: Statue of the drummer on the background of the red building in Amritsar, India (Copyright iStock by Getty Images/credit: Pavel Sipachev and Serhii Brodin).

Sikhism religion and philosophy sprang from Hinduism and Islam in 1469 by Guru Nanak in the Punjab region of the Indian subcontinent. Sikhism members are called Sikhs (Punjabi: disciples or followers). The Sikhs call their belief Gurmat (Punjabi: the way of the guru). Guru is a combination of two Punjabi words; *gu* means darkness and *ru* means light; guru means a spiritual teacher.

Sikhism was founded by Guru Nanak (1469–1539), and the teaching was revealed by God first to Nanak and continued in succession of nine other gurus. All ten gurus that shaped the core beliefs of the religion lived in Punjab between the years of 1469 and 1708 CE.

The Sikhs believe the ten gurus were inhabited by a single spirit. After the death of the tenth guru, Guru Gobind, the spirit of perpetual gurus conveyed the sacred scripture into a holy book called Guru Granth Sahib. Guru Granth Sahib, also known as the Adi Granth (Panjabi: First Book), is a collection of nearly six thousand hymns written by the ten gurus.[118] The Sikhism core belief is one God (Panjabi: Waheguru) and one humanity; the Sikhs' God is the same as the God the Muslims and other monotheistic religions worship.

Leaders

In Sikhism, the most influential leaders are the ten gurus:

1. Guru Nanak
2. Guru Angad
3. Guru Amr Das
4. Guru Ram Das
5. Guru Arjan
6. Guru Har Gobind
7. Guru Har Rai
8. Guru Har Krishan
9. Guru Tegh Bahadur
10. Guru Gobind Singh

Also, there are a few contemporary spiritual leaders. The most important ones are as follows:

Guru Nanak, also known as Guru Nanak Dev Ji, the founder of Sikhism, was born in 1469 CE in the village of Talwandi in the Punjab region of the Indian subcontinent. There is not much information about Guru Nanak's childhood life. He was a member of the trading caste and was working as an accountant, like his father, and he was always composing hymns. At the age of thirty, he had a mystical encounter with God and quit his job and began traveling around India and other countries with his two childhood friends, Bhai Bala and Bhai Mardana. He declared he is neither Hindu nor Muslim, there is only one God, and all human beings have direct access to this God without the need for priests and rituals. He also denounced the caste system and converting Hindus to Islam with a sword; he believed everyone is equal, regardless of gender or caste.

He traveled in Sri Lanka, deep in the Himalayas, Baghdad, Mecca, and Medina to enlighten people with his teaching. Along this path, Muslims and Hindus were converting to Sikhism.

Finally, he settled in Kartarpur, a village on the right bank of the Ravi River in the Panjab, Pakistan. Guru Nanak elected Bhai Lehna as the successor guru. He renamed him Guru Angad (English: part of you or one's very own) and built public meeting temples for Sikhs called gurdwara (English: the door or the gateway to the guru).[119] He died and was buried on September 22, 1539, in Gurdwara Darbar Sahib Kartarpur in Pakistan.

Guru Angad was born in 1504 to Hindu parents and worshipped the Hindu goddess Durga. He led a party to the holy site of Javalamukhi; as he was traveling to Kartarpur, he listened to Nanak's hymns and instantly converted to Sikhism. He was very loyal to Guru Nanak, and therefore, Guru Nanak declared him as the successor guru. He was a successful leader and died in 1552.

Guru Amar Das (1479–1574), who was devoted to the worship of the god Vishnu, spent his life looking for gurus. While he was living along the Ganges River, he became a Sikh. He collected a scripture of hymns of Nanak and added some of his own, called Goindval Pothis. He reestablished rituals to confirm the Sikhs in their faith. He developed the community kitchen (*langar*) to remove the caste contrast and begin social harmony among Sikhs. He ordered the digging of a sacred well and assigned it as a pilgrimage site and, in the meantime, created three festival days that were already Hindu festivals: Magi, Baisakhi, and Diwali.

Guru Ram Das (1534–1581) was the fourth guru and son-in-law of Guru Amar Das. He is the founder of the town of Amritsar (or Ramdaspur), which became the capital of the Sikh religion. This town later became the location of the Harmandir Sahib, or Golden Temple, the place of worship for Sikhs. He wrote the Sikh wedding hymn. Guru Ram Das appointed his son as the successor of the guru, and all the future gurus were from his direct lineage.

Guru Arjan (1563–1606) was the youngest son of Guru Ram Das, who appointed him as his successor. He faced a distinct hostility of his older brother Prithi Chand for his appointment by his father. He collected the most crucial part of the Guru Granth Sahib and completed the building of the Golden Temple. The Mogul Empire was worried about the growth of the Sikh religion, and Emperor Jahangir worried that the Sikh would support Jahangir's rebellious son Khusro. The emperor arrested Guru Arjan and tortured him to death.

Guru Hargobind (1595–1644) was appointed by his father, Guru Arjan. Guru Hargobind changed the Sikhs' Panthan or Panth (English: path) to a way that was either religious and secular, or non-religious. Later, his father deemed him Miri/Piri; *mir* means chieftain of his people to protect them against other forces, and *pir* means religious or spiritual leader. The Mogul Empire imprisoned him for a short time. He completed the dress code started by his father and introduced wearing two swords, one for religious authority and one for his political power.

Guru Har Rai (1630–1661), grandson of Guru Hargobind, had a peaceful time for a while. He moved from Kiratpur and settled with a small entourage at Simur, in Punjab. Occasionally, he went to the plains of Punjab to preach to the Sikhs. Guru Har Rai supported the older brother of Emperor Aurangzeb, Dara Shikoh. Aurangzeb, in response, held his son Ram Rai hostage. Ram Rai worked in favor of Aurangzeb, and he gave him a statesman position. Guru Har Rai opened hospitals to provide free treatment.

Guru Har Krishan (1656–1664) was the second son of Guru Har Rai and succeeded his father at the age of five. While his older brother, Ram Rai, was still being held hostage by Emperor Aurangzeb, the emperor summoned Guru Har Krishan to Delhi from Simur. In Delhi, he contracted smallpox and died. Before his death, he mentioned his successor's name: Tegh Bahadur.

Guru Tegh Bahadur (1621–1675) was the second son of Guru Hargobind. After the news of the death of Guru Har Krishan, many hopefuls rushed to grab the title, but Guru Tegh Bahadur succeeded Guru Har Krishan. His original name was Tyag Mal, but his subjects named him Tegh Bahadur (English: brave sword) because of his resistance to Emperor Aurangzeb. He wrote several hymns that are recorded in the Guru Granth Sahib. He predicted the downfall of the Mogul Dynasty, and Emperor Aurangzeb gave him two options: death, or conversion to Islam. He accepted death, and he was beheaded and cremated.

Guru Gobind Singh (1675–1708) became the spiritual and temporal leader of the Sikhs at the age of nine, and the founder of the Khalsa[120] (English: the brotherhood of the pure). His original name was Guru Gobind Rai; after the creation of Khalsa, he was renamed Guru Gobind Singh. The surname Singh is derived from Sanskrit Simha, meaning lion or brave. At the age of five, he was educated in Sanskrit and Persian. He was the most famous, after Guru Nanak, as a warrior, a prophet, and a poet. Guru Gobind Singh gave all Sikh men the surname Singh and all women the surname Kaur, meaning princess. Sikhs came to have a warrior attitude because of atrocity and violence against them by the Islamic empire of the Moguls. In response, Guru Gobind Singh introduced the Khalsa to the Sikhs to wear the following, called the five Ks:

1. Kes or Kesh means uncut hair.
2. Kangha means comb.
3. Kachha means short trousers.
4. Kara means steel bracelet.
5. Kirpan means ceremonial sword.

Sikhs believed they were justified in drawing the sword. Guru Gobind Singh expressed his belief in Zafar-Nama, indicating the Epistle of Victory, a letter sent to Emperor Aurangzeb. He also declared that the scripture of the Guru Granth Sahib is the authority from which the Sikhs would be governed. If the scripture had to be moved, it must be driven by five Sikhs who represent the Khalsa. After several attacks by other Shiwalik chieftains in cooperation with the governor of Sirhind, he lost his two sons and finally was assassinated.

This movement under the name of religion started with Guru Nanak to fight against the caste system of the Hindu religion, as well as the cruelty of Islam during the Arabs' invasion, forcing Hindus to convert to Islam and spread Islam in India. It can be seen from the lives of the ten gurus the cruelty of Muslims for fear of their downfall. In the end, Sikhism continues to exist as a small religion, not as a fighting power against Hindus' inequality nor Muslims' atrocities.

Scriptures

Adi Granth (Panjabi: First Book) is the sacred scripture of Sikhism. The fifth guru or Guru Arjan compiled the Adi Granth (also called the Guru Granth Sahib). The compilation started in 1601 and was completed in 1604. It is a collection of hymns from Gurus Nanak, Angad, Amar Das, and Ram Das, with a group of devotional songs from Hindu and Islamic saints, especially the poet Kabir. The Adi Granth is placed in the Harmandir Sahib (House of God) with a huge celebration.

In 1704, Guru Gobind Singh added the teaching and hymns of his predecessor, Guru Tegh Bahadur (Guru Hargobind, Guru Har Rai, and Guru Har Krishan did not write hymns). He commanded that after his death, the Guru Granth Sahib will take the place of the final guru. The Guru Granth Sahib contains about 6,000 hymns in 1,430 pages. The hymns are organized by the musical notes in which they are to be sung. Every gurdwara has a central copy of Guru Granth Sahib.

After the death of Guru Gobind Singh in 1708, his large number of hymns and other writings were compiled into another book, known as the Dasam Granth. Some poems and writing could be from other sources. Some of the hymns from the Dasam Granth are used in Sikh rituals and worship. By no means does Dasam Granth reach the level of reverence to the primary scripture from the Guru Granth Sahib. Sikhs have more scriptures, but those will not be discussed in this book.[121]

Denominations

Like every other religion, in Sikhism, too, individuals came up with some scheme to have control over people. In the over five hundred years of existence of Sikhism, several denominations and groups have broken away from orthodox Sikhism.

1. Nirankari is based on the teaching of Baba Dyal (Baba Dayal Das). It was founded by Gautam Singh.
2. Nam-Dharis (Kuka Sikhs) was founded by Balak Singh, who introduced himself as the successor of Guru Gobind Singh.
3. Randhir Singh founded Akhand Kirtani Jatha.
4. Guru Gobind Singh initiated Nirmala.
5. Shiv Dayal Singh Seth founded Radha Soami in 1869.
6. Udasi was founded by Baba Siri Chand, the eldest son of Guru Nanak, because Guru Nanak elected Bhai Lehna as his successor guru.

Beliefs

Sikhs believe in only one omnipotent and omnipresent being or Almighty God or Great Teacher (Panjabi: Waheguru) that is formless and genderless. Sikhs believe Almighty God existed alone before the creation of the universe. He created the universe, including all living beings, and he is present in everything. Sikhs believe all people are equal regardless of race and gender, and the believers communicate with Almighty God through faith and meditation. Sikhs follow the teaching of the ten gurus; the first, Guru Nanak, and the last, Guru Gobid Singh, who declared that the Sikh sacred scripture is the final eternal Guru, and that is the Guru Granth Sahib. Some of the extracts from the holy scripture of the Guru Granth Sahib are as follows (which are common in the Hindu religion or against Muslims' atrocity):

Karma and Mukti: Sikhism teaches that human beings spend their lives in the cycle of birth, life, and rebirth, and share this concept with slight modification with followers of other Indian religions like Hinduism, Jainism, and Buddhism. Guru Nanak says in the sacred scripture of the Guru Granth Sahib that the human body takes birth because of karma, but liberation is achieved through God's grace. Mukti means freedom is the process of replacing ignorance and ego by spiritual enlightenment touched by God's grace or the gift of God, ultimately becoming one with God, free from the cycle of life and death. Mukti is similar to Hinduism's moksha as far as liberation.

Hukam: Hukam is a Panjabi word derived from the Persian and Arabic *hukm*. It means divine will, order, or command. Sikhs have to live in coherence with hukam to attain salvation and avoid selfishness and ego. As Guru Nanak says, "What pleases Thee, O Lord, that is acceptable. To Thy Will, I am a sacrifice" (AG, p. 685).

Haumai: Haumai is a combination of two Panjabi words, *hau* meaning I, and *mai* meaning me or simply ego, self-centeredness, or false self-representation. The person who wants to satisfy their ego or haumai must follow their desires, what they want. Ego or haumai leads a person into *dukh* or hardship, ailment, and suffering. Haumai can be overcome through meditation on God's names.

Nimrata: Nimrata, a Panjabi word (English: humility or benevolence), is a virtue considered the opposite of Haumai, which is promoted by the Guru Granth Sahib. Sikhs believe meditation on God and selfless service cultivate Nimrata and eliminate Haumai.

Seva: Seva is a Panjabi word (English: service). Sikhs serve God by serving other people and their community (or other communities) every day. By serving and devoting their lives to helping others, they get rid of their selfishness and ego. Sikhism does not tell Sikhs to ignore their own life to get closer to God. Many Sikhs perform tasks in the gurdwara, like washing dishes, cleaning the floors, and working in the kitchen. Sikhs have a tradition of seva, or service. A person who performs seva through service, voluntary or altruistic, is called a *sevada*.

The Three Duties: Guru Nanak developed, taught, and directed the Sikhs to carry out the Three Duties in life as follows:

1. *Naam Japna* (or Naam Simran) is the meditational chanting of the Waheguru, which means Wonderful Lord.
2. *Kirat Karo* means earning an honest living. Sikhs try to live honestly because God is honest. Sikhs do not beg, gamble, smoke, drink, or work in the tobacco or alcohol industries.
3. *Vand Chakko* means to share the fruits of one's labor with others before considering oneself, and giving to charity and caring for those who are in need.

The Five Vices within the Human Body: These five vices are the great enemy of humans and causes of suffering.

1. *Lust* is sinful, and therefore, Sikhs wear short trousers to remind them of the shame and misery of nonmarital sex.
2. *Greed* is the desire for wealth and power. Often, greed causes fraud and immorality.
3. *Attachment or Moh* means excessive love attachment of a wife or material goods and things of the world.
4. *Anger* is a feeling of eagerness and enthusiasm that could cause quarrels and violence. Anger is overcome by tolerance.
5. *Ego or pride* is the worst of all—the remedy for ego is humility.

Afterlife: Sikhs believe in the reincarnation, karma, or intentional action, which agrees with Hinduism, Jainism, and Buddhism. Sikhs do not believe in hell or heaven. Hell and heaven can be experienced on the earth while alive. Suffering and pain caused by ego are seen as hell, and being in tune with God is seen as heaven. Death is a very short ceremony; the dead will be washed, placed in clean cloth, and decorated with the Five Ks; close relatives take the deceased for cremation. The ashes will be disposed of in the nearest river. There are other methods of burial, like in-ground or in the sea, but these are very uncommon.

Rituals and Customs

Guru Nanak declared that he is neither Hindu nor Muslim, there is only one God, and all human beings have direct access to this God without the need for priests and rituals. Based on this statement, Sikhism has no priestly class, no sacraments, and very few ceremonies.

Gurdwara (Dharamsala): The gurdwara (English: the door or the gateway to the guru) is a place for Sikhs to worship. In a gurdwara are four doors, called the door of peace, the door of learning, the door of livelihood, and the door of grace. The doors are a symbol that people from all four points of the compass are welcome, and any caste member is equally welcome. The gurdwara is a simple structure that contains a cot under a canopy with a copy of the Adi Granth, the first volume of the sacred book of Sikhs. At the center of the gurdwara is the spacious hall where the copy of the Sri Guru Granth Sahib is located for devotional activities and recitation of hymns. The location of the Sri Guru Granth Sahib is called Darbar Sahib. Sikhs enter this hall the same way as they enter a mosque; they take their shoes off and pay a couple of pennies. The gurdwara also serves as the meeting place for weddings, initiation ceremonies, and langar (the soup kitchen).

Langar: Langar, or the free-food kitchen, is attached to every gurdwara. This is where food is served without charge regardless of religion, caste, or gender, and all social castes eat together on the floor. Guru Nanak started this tradition with vegetarian food. At the same time, Sikhs are not vegetarians to be inexpensive, but to prevent rich and wealthy congregations from turning this concept into a feast and showing off their dominance. One type of food that is popular and Sikhs share is called halvah, and is made of crushed sesame seeds in a syrup, such as honey or clarified butter and sugar.[122]

Santsang: Santsang means religious gathering, or the assembly of true believers. This practice goes back to Guru Nanak. Sikhs listen and sing hymns and compositions from the Sri Guru Granth Sahib in the gurdwara and share meals. This gathering is open to everybody.

Granthi and sevadar: Granthi comes from Granth, the holy scripture. A granthi is a person who deals with all aspects of the Sri Guru Granth Sahib, like installation every morning at dawn, closure at dusk, reading, teaching, interpretation, propagation of its messages to the congregation, leading prayer, leading the congregation in worship, and taking care of the gurdwara. In short, a granthi is a spiritual and religious leader. In addition to granthi, in Sikhism the person who volunteers to perform seva, or selfless service without compensation in the gurdwara and langar facility is called a *sevadar*. Sevadars follow the Three Duties of Sikhism.

Worship: Sikhs can pray in public or private, any time, and any place. They get up early, bathe, and then start the day with Naam Japna, or the meditational chanting of the Waheguru (Wonderful Lord). They pray three times each day: in the morning at daybreak, in the evening at sunset, and before going to bed. Sikhs recite five *gurbani*, or five prayers, which are passages from the Sri Guru Granth Sahib; they are called nitnem, or daily habits, written in Gurmukhi script. The morning nitnem are five gurbani, evening has one, and bedtime has one.

Marriage: The Sikh marriage is known as Anand Karaj, which was introduced from the Sikh gurus and was given official statutory recognition by the British under the Anand Marriage Act of 1909. Under the Sikh Rehat Maryada, or central approved Sikh code, the bride and groom not professing the Sikh faith cannot be married in the Anand Karaj ceremony. The ceremony can take place in a gurdwara or any other place with an installation of the Sri Guru Granth Sahib. The groom meets the family of the bride, and the two families recite the Ardas prayer; the groom brings a silk offering for the Sri Guru Granth Sahib. Then the bride and groom sit, and the officiating individual (who could be a granthi or a Sikh man or woman) reads the four Lavan, or stanzas, from the Sri Guru Granth Sahib. After the reading of the first stanza, the bride and groom rise and slowly walking around the Sri Guru Granth Sahib clockwise until completion. All invitees then stand to recite the Ardas, and halvah or

Karah Prashad[123] is distributed to end the ceremony. The following are rules in Sikhs marriage: child marriage is prohibited, all Hindu superstitions about good days or bad days are forbidden, Sikhs practice only monogamy, and widows or widowers can remarry.

Music: Sikh music (Shabad kirtan or Shabad keertan) started with the rise of Sikhism as the musical expression utterance of mystical poetry devised by Guru Nanak.

Bhai Mardana, a Muslim who was the son of a Mirasi couple, was a professional rababi player (a rabab is a musical instrument). Guru Nanak and Bhai Mardana were childhood friends. They traveled for many years together; Guru Nanak composed hymns and sang while Bhai Mardana played the rabab with him in the Kirtan-style singing of hymns.[124] Following Guru Nanak, all the Sikh gurus sang in classical and folk music. The songs are hymns from the Sri Guru Granth Sahib. Music plays a significant role in the life of Sikhs.

Amrit Sanchar: Amrit Sanchar (Khanda-Ki-Pahul), also called Amrit Parchar, is the Sikh baptism ceremony. This ceremony was established in 1699 by the final and tenth guru leader Guru Gobind Singh when he founded the Khalsa. His five faith followers were baptized first, and then the five faith followers (or Panj Pyara) baptized him. A Sikh boy and girl can go through this initiation when they reach adulthood. The ceremony takes place in the presence of five initiated Sikhs in a gurdwara before the Sri Guru Granth Sahib. They drink a mixture of water, sugar, and tukmaria (basil) called Amrit. This drink will also be sprinkled on their eyes and hair.

Festivals and Holy Days

Sikhism is based on the spiritual teaching of Guru Nanak and ten successive Sikh gurus, and became popular with Hindus and some Muslims. Sikhs use some Hindu festivals, but none from Muslims. The Sikhs festivals are as follows:

Parkash Utsav Dasveh Patshah (January 5): The name of this festival translates to Birth Celebration of the Tenth Divine Light. It celebrates the birth of Guru Gobind Singh, the tenth and last human guru. This festival is one of the most important events celebrated by Sikhs.

Maghi (January 14): This festival commemorates the Battle of Muktsar, during which some forty soldiers of the Mogul army deserted and joined Guru Gobind Singh. They fought bravely against the Mogul army and were martyred in Muktsar. Guru Gobind Singh blessed them as mukti, or liberation, and cremated them at Muktsar. This festival was chosen by Sri Guru Amar Das for Sikhs to celebrate in the gurdwara, and also coincides with Makar Sankranti, which is a Hindu winter festival.

Hola Maholla (March 26): Hola Maholla is a one-day mela,[125] or festival, celebrated in Anandpur Sahib, which coincides with the Indian festival of color. Guru Gobind Sahib instituted Hola Maholla as a gathering of Sikhs for military exercises and mock battles.

Vaisakhi (April 14): This festival is a celebration of the birth of the Khalsa brotherhood in 1699 which gave Sikhs their identity. Sikh enthusiasts attend the gurdwara before dawn with the offering, and some will be baptized.

Martyrdom of Guru Arjan (June 16): This celebrates the martyrdom anniversary of Guru Arjan, the fifth guru, who was tortured and killed under the order of Jahangir, the Mogul emperor in 1606. Sikhs gather at the gurdwara and happily accept his torture and death as a will of Waheguru.

Pahila Prakash Sri Guru Granth Sahib Ji (August or September): This festival celebrates when the Sri Guru Granth Sahib was completed in 1606 and the end of the human gurus.

Bandi Chhor Diwas (October or November): Bandi Chhor Diwas coincides with the Hindu Festival of Lights (Diwali). Sikhs celebrate the release from prison of Gurur Hargobind, the sixth guru, who also rescued fifty-two Hindu kings kept captive by Jahangir, the Mogul emperor, at Gwalior Fort in 1619. The Golden Temple is illuminated, and Sikhs decorate their homes with candles and attend the gurdwara.

Guru Nanak Gurpurab's Birthday (November 15): Guru Nanak was born on November 15, 1469, in Nanakana, Sahib, Pakistan. On this day, Sikhs celebrate by lighting candles and burning fireworks in honor of Guru Nanak. Two days before his birthday, Sikhs gather in the gurdwara for forty-eight hours of nonstop reading of the Sri Guru Granth Sahib.

Martyrdom of Guru Tegh Bahadur (November 22): On this day in 1675, Guru Tegh Bahadur, the ninth guru, was executed by Emperor Aurangzeb.

Summary

Sikhism is one of the religions specific to the Indian subcontinent. The founder of Sikhism was Guru Nanak (1469–1539). Sikhism is a monotheistic religion, with the core belief of one God (Panjabi: Waheguru) and one humanity. The Sikh God is the same God as Muslims and other monotheistic religions worship. The emergence of Sikhism was a reaction against the teaching of Orthodox Brahmanism with the caste structure and the ferocity of Muslims killing Hindus to force them to convert to Islam.

In Sikhism, like every other significant religion, fear is one primary component.

Karma: Sikhs believe in the incarnation, karma, or intentional action, which agrees with Hinduism, Jainism, and Buddhism. Sikhs do not believe in hell or heaven. Hell and heaven can be experienced on the earth while alive. Suffering and pain caused by ego are seen as hell, and being in tune with God is seen as heaven.

CHAPTER 15

Baha'i

Baha'i Temple in Haifa Mount Carmel, Israel (Copyright iStock by Getty Images/credit: Adam Keinan and Grebeshkovemaxim).

Baha'i, also called the Baha'i Faith or Baha'ism, is a religion founded in Iran in 1863 by Mirza Hosayn Ali Nuri, who is well-known as Baha-Allah or Baha-Ullah. Baha-Ullah is an Arabic name meaning Glory of God, and in this book, this religion will be called Baha'i. The Baha'i religion is the continuation of Babism, which arose from Shia Islam.

Babism was originally founded in 1844 by Sayyed Mirza Ali Muhammad Shirazi of Shiraz, Iran. He declared a new spiritual doctrine that affirmed the progressiveness of revelation, stating that no revelation was final. He forbade polygamy and concubinage, solicitation, the use of intoxicating material, drugs, and dealing in slaves. The word *bab* in Persian means gateway. He proclaimed that a new prophet (messenger of God) would come who would rescind the old beliefs and guide new beliefs. His doctrine already existed in Shia Islam. Shia believed in the coming of the twelfth imam, Muhammad ibn Hassan Al-Mahdi, the successor of Muhammad, who guides the loyal believers.

Sayyed Mirza Ali Muhammad Shirazi's teaching spread throughout Iran, and this new religion was a severe danger to the very existence of Shiite Muslim clergy and government. He was arrested, confined in prison for several years, and executed in 1850; twenty thousand of his followers were killed, and a large number of his followers left Iran for India, Iraq, and other countries.

Soon after the death of Sayyed Mirza Ali Muhammad Shirazi, his disciple Mirza Hosayn Ali Nuri, known as Baha-Ullah, joined the Babis. In 1852, he was arrested and confined in prison, where he proclaimed that he got a revelation from God and was assigned to be the prophet and messenger. In 1853, he was released from prison and exiled to Baghdad. In 1863, he declared to the Babist community that he was the messenger of God as the Bab foretold. Babists acknowledged him, and from that point on, he became known as Baha'i. In 1863, he was sent to Constantinople (now Istanbul), Turkey, by the Ottoman Empire and confined. Later, he was imprisoned in Adrianopole (currently Edrine), Turkey, and finally ended up in Akko, Palestine, now Acre, Israel.

Baha-Ullah died in 1892; before his death, he appointed his eldest son, Abdul Baha, to be the leader of the Baha'i religion and the primary interpreter of his teaching and script. After obtaining relief from exile in 1908, Abdul Baha began traveling to North America, Europe, and other continents to spread his father's teaching and messages. Abdul Baha appointed his eldest son, Shoghi Effendi Rabbani, as his successor.

Baha'i has its own calendar. It is a solar calendar with years composed of nineteen months of nineteen days in each month, a total of 361 days, plus extra days of intercalary days. The New Year begins on March 21 and is called Naw-Ruz; it goes back to 2000 BC in Khorasan.

Leaders

Baha'is have only a few leaders, and some of them were executed and exiled.

The Bab, or Sayyed Mirza Ali Muhammad Shirazi: Sayyed Mirza Ali Muhammad Shirazi was born on October 20, 1817, in Shiraz, Iran, to a middle-class merchant family of Shia Muslims. He was given a Shia birth name of Ali Muhammad. In 1842, he married Khadijih-Sultan Shirazi (Khadija Khanum). Ali Muhammad traveled on a pilgrimage to Mecca, Medina, and stayed for a couple of months in Karbala, where he met some Shaykhis of the 1790s movement in Iran. The Shaykh movement was expecting the forthcoming appearance of the imam Al-Mahdi (Al-Qaim). He met Kazim Rashti, the leader of the Shaykhis movement. In 1844, Ali Muhammad declared himself to be the Al-Mahdi of Hidden Imam and gave himself the title of the Bab.

Babism spread all over Iran, and he amassed thousands of followers. The religion of Babism was a danger to the very existence of Shiite Muslims, clergy, and government. He was arrested and confined in prison for several years in different provinces of Iran and executed in 1850 after a short trial for heresy against Islam. Twenty thousand of his followers killed, and a large number of his followers left Iran for India, Iraq, and other countries. Bab tried to follow what Muhammad followed, to no avail. Muhammad created an army from bounty raids and looting of caravans and later, slaughtering Jewish communities and grasping their wealth and land. Gradually, without resistance, he created an empire. Bab's time was different; the opposition force was powerful, and they were able to crush him easily.

Baha-Ullah (Mirza Hosayn Ali Nuri): Mirza Hosayn Ali Nuri was born on November 12, 1819, in Tehran, Iran, to a merchant family of Shia Muslims. He died on May 29, 1892, while he was living in confinement in exile in Acre, Israel. The Acre became a town of pilgrimage for Baha'i believers.

Mirza Hosayn Ali Nuri was one of the disciples and followers of the Bab, who experienced a revelation from God in 1852 that he was the divine messenger prophesied by Bab. In 1863, he publicly proclaimed himself Baha-Ullah, meaning Glory of God to be the imam Al-Mahdi, or Hidden Imam (rightly guided leader). While he was in Acre, Israel, he began writing Baha'i doctrine that recommended the unity of all religions. Before his death, he named his son Abdul-Baha as his successor.

Abdul-Baha (Abbas Effendi): Abbas Effendi, son of Baha-Ullah, was born on May 23, 1844, in Tehran, Iran. He died on November 28, 1921, in Haifa, Israel, and is buried in the Shrine of Abdul-Baha. He changed his name to Abdul-Baha, which means servant of Baha. At the time Abdul-Baha was in exile, he administered the construction of the first Baha'i house of worship in the city of Ishqabad, Turkmenistan, and the Shrine of the Bab in Haifa, where the Bab is buried. Also, he administered the restoration of the House of the Bab in Shiraz, Iran. In 1908, after the Young Turk Revolution, he left Acre to travel around the world and spread Baha'i teaching. Prior to his death, he named his grandson Shoghi Effendi as his successor.

Shoghi Effendi: Shoghi Effendi[126] was born on March 1, 1897, in Acre, Israel, and died on November 4, 1957, in London. He is buried in Southgate Cemetery and Crematorium in London. In 1918, he earned his bachelor of arts degree from the American University in Beirut, Lebanon. He further studied at the Balliol College in Oxford, United Kingdom. He returned to Haifa in 1921 to undertake the office of the Guardian, given to him as a result of the unexpected death of his grandfather. He was serving as a translator of the teaching of his grandfather Abdul-Baha and great-grandfather

Baha-Ullah, and the Bab to English. During his guidance, the Baha'i Faith quadrupled in number as Effendi created spiritual assemblies to guide the Baha'i Faith to other parts of the world. In 1937, he started a series of plans to establish Baha'i communities worldwide. Also, a Ten-Year Crusade was expected from 1953 to 1963, with creating the Universal House of Justice as its supreme goal. After Shoghi Effendi's death, Baha'i leadership passed on to the Universal House of Justice.

Scriptures

Baha'is have tolerance toward other religions. They respect their prophets as divine messengers and respect their sacred scriptures, mainly Islam, Judaism, Christianity, Zoroastrianism, Hinduism, and Buddhism. Baha'is view the writing of the Bab and Baha-Ullah as divine revelations. Baha-Ullah's writings are more than a hundred books and tablets. Most of his writings were written under the difficult environments of imprisonment. The primary Baha'i scriptures are the following:

- Kitab-i-Asma (Book of Divine Names)
- Dalail-i-Sabih (the Seven Proofs)
- Kitab-i-Aqdas (the Most Holy Book)
- Kitab-i-Iqan (the Book of Certitude)
- Kalimat-iMaknunih (the Hidden Words)
- Haft-Vadi (the Seven Villages)
- Char-Vadi (the Four Valleys)
- Kitab-i-Ahd (the Book of the Covenant)
- Gleanings from the Writing of Baha-Ullah
- The Summons of the Lord of the Hosts
- The Tabernacle of the Unity
- The Tablets of Baha-Ullah

Denominations

The Bab and Baha-Ullah were born Shia Muslims in a Shia Muslim society in 1249 and 1247, respectively. Both tried to be a leader like Muhammad and manipulate others with their new ideas. During the rise of Muhammad, there was no significant power to crush Muhammad's growth, while during the rise of the Bab and Baha-Ullah, the opposition forces were powerful, and they were able to smash them. Bab was killed, as well as twenty thousand of his followers, and Baha-Ullah was exiled until his death.

The total population of the Baha'i Faith is about six million.[127] Despite the small number of believers and less than two hundred years of Baha'i religion, there are approximately nine denominations, with minor differences:[128]

1. Orthodox Baha'i Faith: The Orthodox Baha'i Faith following Joel B. Marangella.
2. Baha'is Under Provisions of the Covenant: This denomination follows Leland Jensen.
3. Tarbiyat Baha'i Community: This denomination, formerly known as Orthodox Baha'i, is under the leadership of Rex King.
4. Baha'is Under the Living Guardianship: This denomination follows Donald Harvey as the third Guardian and Jacques Sonhomonian as the fourth Guardian.
5. Reform Baha'is.
6. Free Baha'is.
7. Unitarian Baha'is.
8. The Baha'i Faith or Baha'i World Faith.
9. John Carre: This denomination follows Alif, a Third Manifestation of God—the third letter of the Greatest Name, the Spiritual Foundation of the Society.

Beliefs

The Baha'i founder, Baha-Ullah, was born as a Shia Muslim, and the idea of the forthcoming appearance of the Al-Mahdi (messenger of God) had an imprint in his brain that was reinforced by Babism and the Shaykhi movement. The foundation of Baha'i is Shia Islam, and also some of his ancestral religion of Zoroastrianism and, later, Christianity. Baha'i believe that this age is leading up to a final unity of all people and beliefs of the earth, with one language.

Creation: Baha'i has no clear view about creation and offers two viewpoints. On one hand, they believe that the universe is eternal and it has always existed, like the creator; on the other hand, they think that individual elements of the earth came into being in a specific time, and these elements cease at some point.

Deity: Baha'is believe there is only one God, the creator of the universe, and God is an imperishable, all-knowing, uncreated being, unknowable and inaccessible. God is the source of revelation, and is eternal, almighty, omnipresent, and omnipotent. God sends prophets and messengers at a particular time, place, and level of human understanding.

Prophets: In the Baha'i belief, God sends messengers or prophets to provide the most absolute and thorough knowledge of God. These prophets and messengers are the manifestations of God. The writings and scriptures created by these prophets or messengers are how people get a better knowledge of God and help to shape society. The manifestation of God is in Adam, Abraham, Moses, Krishna, Zoroaster, Buddha, Jesus Christ, Muhammad, Bab, and the Baha-Ullah. Baha'is believe that the messengers provide knowledge of God as a function of time in a progressive way, and a new messenger provides up-to-date knowledge of God.

Scriptures: All the writing of the Bab and Baha-Ullah are regarded as godly revelation. The scriptures of other faiths, like Buddhism, Judaism, Christianity, and Islam, are regarded as divine disclosure and the teachings of previous manifestations of God. The writing of Abdul-Baha and Shoghi Effendi is a canonical interpretation of the Bab and Baha-Ullah's godly revelation.

Worship: The main goal of Baha'i life is to know and love God. To achieve this goal they perform prayer, fasting, and meditation. Baha'is believe service to the people is a kind of worship. Their mission is to bring unity and harmony to the world, and therefore, they meet in community and monthly feasts to build unity and harmony. These gatherings are called Nineteen-Day Feasts, and start before the first day of each Baha'i month.

Satan: The Baha'i Faith and teaching of the Baha-Ullah deny the existence of Satan (Arabic: Iblis or shaitan). In the Abrahamic religions, Satan was an angel who acted against the will of God. In the same way, the Baha'i Faith and teaching of the Baha-Ullah deny the existence of heaven and hell. Baha'is believe heaven is the natural outcome of spiritual progress, and hell is the result of failure to progress spiritually.

The soul: Human life begins when the soul connects with the embryo at the very moment of conception. It is the soul that gives humans the power to think, to understand, to live, and achieve progress. When death happens, the soul is separated from the dead body and continues the eternal journey and unites with God.

Salvation: Unlike the Abrahamic religions, which believe in salvation from original sin and other evil actions in the next life, Baha'is understand salvation from self-centeredness and captivity in despair in this life. Baha-Ullah expressed that pride or self-centeredness is one of the biggest impediments in human life. A self-centered person feels that they are in complete control not only of their own life, but also of their surroundings. They seek dominance over others because such dominance helps them carry on this illusion of superiority. A person can achieve salvation by relying upon the teachings of Baha-Ullah as a mediator between humanity and God to overcome their self-centeredness and other disastrous illusions.

Funeral: Baha'is believe that upon the death of a person, their soul is released from its ties with the physical body and enters the eternal spiritual world. Baha'is focus on comforting the living and blessing the departed soul. According to Baha'i law, the deceased must be buried less than an hour's journey from the place where he or she died. The burial must happen quickly, within two to three days after death. The position of the deceased in the grave is with feet pointed toward Qiblah,[129] where Baha-Ullah is buried. The funeral service tends to be very simple; all recite the prayer of death, and there are no condolences or food offerings.

Death and afterlife: Baha'is believe in life after death, and Baha'i teaching affirms that the human soul lives forever along a path toward perfection, joy, and unity with God. Baha-Ullah declared that we cannot understand what the next life after death will be like any more than the child in the womb can conceive of life in this world. We have no possible frame of reference.

Prohibitions and permissions: Prohibitions and permissions are according to the Baha'i laws, instructed by Baha-Ullah and written in the Kitab-i-Aqdas.[130] There is a long series of dos and don'ts. Don'ts are the following:

- Avoid backbiting and gossip.
- Avoid alcohol, tobacco, and mind-altering drugs.
- Polygamy is not allowed.
- No sexual relationship outside marriage.
- No interpreting the Baha'i writings.
- Slavery is not allowed.
- No asceticism.
- No monasticism.
- No begging.
- No clergy.

The dos are the following:

- Marriage is recommended but not obligatory.
- Interracial marriages are encouraged.
- Pilgrimage.
- Fasting for people aged fifteen to seventy from March 2 to March 20.
- Daily prayer between the ages of fifteen to seventy, facing Qiblah, where Baha-Ullah is buried.
- Working for the welfare of society.
- Promoting the unity of humankind.

Rituals and Customs

The Baha'i Faith, in comparison to other religions, has minimal laws and rituals. All the activities and practices are guided more by moral principles than statutes. They do not have clergies or initiation ceremonies.

Prayers: Baha-Ullah directed Baha'is that they should pray every day. He gave them three prayers, and they can select which one to use daily. One prayer is very short, and can be said between noon and sunset; the next one is a medium-length prayer, and can be said three times a day; the third one is a long prayer, and can be said once daily any time of the day.

Fasting: Baha'is aged fifteen to seventy must fast (like Muslims) nineteen days a year from March 2 to March 20. Baha'is who fast must avoid eating, drinking, sexual activity, smoking, and gambling, from sunrise to sunset. Excused from fasting are those who are ill, traveling for more than one hour, pregnant, and nursing mothers.

Marriage: Marriage is not obligatory but is strongly recommended. The union of a man and woman as a family is the fundamental building block of communities. Baha'i marriages adhere to the customs of their culture. Both parties must agree to the union, as well as the parents of both sides. During the wedding ceremony, the man and woman must recite a wedding vow from the Kitab-i-Aqdas, and that is "We will all, verily, abide by the Will of God."

Pilgrimage: Baha-Ullah was exiled to Akka, Palestine, in what is now the city of Acre in Israel. While he was alive, his followers visited him there. After his death, he was buried next to his residence. After the Bab's execution, his remains were secretly transferred to many different places. Finally, they buried his remains at the Shrine of the Bab in Haifa, Israel, on the middle terrace of the Baha'i Garden. The city of Acre is the holiest city for Baha'is, and it is a yearly pilgrimage site. These pilgrimages last nine days.

Another Baha'i pilgrimage site is the House of the Bab in Shiraz, Iran, and the House of Baha-Ullah in Baghdad, Iraq. The House of Baha-Ullah in Baghdad was recently destroyed and is no longer a site of pilgrimage for Baha'is.

Festivals and Holy Days

Baha'is celebrate nine festivals to commemorate some special sacred events, including the Nineteen-Day Fast. All the festivals are celebrated following the Baha'i calendar. The Baha'i calendar begins from 1844 CE and uses the solar years of nineteen months and nineteen days in each month. They add four days before the last month of the year to make the number of the days 365, and add five days to the leap year.

Nineteen-Day Fast: This fast starts on March 2 and lasts nineteen days, until March 20, which is the last month of the Baha'i calendar. During this month, Baha'is who fast must avoid eating, drinking, sexual activity, smoking, and gambling, from sunrise to sunset.

Naw-Ruz: Naw-Ruz is on March 21 is the traditional New Year celebration that goes back to 2000 BC in Khorasan. It is the traditional first month of Baha'ism, as well as Zoroastrianism. Baha'is take the day off from school and work to celebrate the New Year.

Ridvan: Ridvan starts on April 21, at sunset; the ninth day is April 29, and it concludes on the twelfth day, May 2, at sunset. Baha'is commemorate Baha-Ullah's declaration that he was the divine messenger prophesied by the Bab, in the Garden of Ridvan, near Baghdad. The first day marks the arrival of Baha-Ullah; the ninth day, the appearance of Baha-Ullah's family; and the twelfth day, their departure from the Garden of Ridvan.

The Bab's Declaration: Baha'is celebrate the Bab's declaration of his mission on May 23, 1844. On this day, he declared that he was the Al-Mahdi (Hidden Imam and God's divine messenger). This day Baha'is stay home from work and school.

Passing of Baha-Ullah: Baha'is observe the anniversary of the death of Baha-Ullah in exile, the founder of Baha'i Faith, on May 29, 1892, at the age of seventy-five. Baha'i schools are closed, and work is suspended.

Martyrdom of the Bab: On this day, Baha'is commemorate the anniversary of the Bab's (Sayyed Mirza Ali Muhammad Shirazi) execution by a firing squad on July 29, 1850, in Tabriz, Iran. On this day, Baha'i schools are closed, and work is suspended.

Twin Holy Birthdays: The Bab was born on October 20, 1819, and Baha-Ullah was born on November 12, 1817. According to the Islamic lunar calendar, both birthdays occurred on consecutive days of Muharram, the first and second. In 2014, the Universal House of Justice resolved to celebrate these days on the first and second days of the eight new lunar months after Naw-Ruz. These days, Baha'i schools are closed and work is suspended.

Day of the Covenant: The Day of the Covenant is celebrated by Baha'is to commemorate Baha-Ullah's nomination of his eldest son, Abdul-Baha, as leader of the Baha'i.

Passing of Abdul-Baha: Baha'is observe the anniversary of the death of Abdul-Baha, son of Baha-Ullah and his appointed successor, on November 28, 1921, in Haifa, Israel.

Summary

Baha'i, also called the Baha'i Faith or Baha'ism, is a religion founded in Iran in 1863 by Mirza Hosayn Ali Nuri, who is well-known as Baha-Allah or Baha-Ullah. The Baha'i religion is the continuation of Babism.

Babism was originally was founded in 1844 by Sayyed Mirza Ali Muhammad Shirazi of Shiraz (the Bab), in Iran. The word *bab* in Persian means gateway. He proclaimed that a new prophet or messenger of God would come who would rescind the old beliefs and guide new beliefs. Shias believe in the coming of the twelfth imam, Muhammad ibn Hassan Al-Mahdi, the successor of Muhammad as Bab claimed.

The Bab was arrested, confined in prison for several years, and executed in 1850. Twenty thousand of his followers were killed, and a large number of his followers left Iran for India, Iraq, and other countries.

Baha'is do not have an opinion about creation, and believe that after death, the soul joins God; there is no heaven and no hell. In Western countries, Baha'i has made some progress, and presently there are around six million Baha'is in the world.

Scientology Cross
(Credit: Ricochet64)

CHAPTER 16

Newer Religions

Pentecostalism
(Credit: Coolvectormaker)

Hare Krishna
(Credit: Coolvectormaker)

Rastafarians
(Credit: Margarita Miller)

Native North American Shaman
(Credit: Ice 245)

1—The Church of Scientology

Scientology International teaches that people are immortal spirits who have forgotten their true nature. Scientology is a philosophical religious movement. The beliefs are a set of instructions written by L. Ron Hubbard (1911–1986), the founder of Scientology, in 1954. Before founding Scientology International, L. Ron Hubbard created a corporation called the Dianetics Foundation in 1950. He published his bestselling book *Dianetics: The Modern Science of Mental Health* through the Dianetics Foundation.

In the beginning, the Dianetics Foundation did very well, and lots of money was pouring in. He applied his made-up mental health from outdated Freudian theory. Those who used it did not see a positive result, and the professional and scientific community strongly condemned its application. Soon, the Dianetics Foundation went bankrupt. L. Ron Hubbard then recharacterized the subject as a religion to fool the average person under the name of religion. In 1954, he renamed the Dianetics Foundation the Church of Scientology. He retained all the terminology and methods used in the Dianetics Foundation, such as the following:

Engram: A memory trace or a hypothetical perpetual change in the brain accounting for the existence of memory. Karl Lashley, a neuropsychologist researcher at Harvard University, proved that the idea of an engram being "a memory repository within the brain"[131] was false.

E-meter: E-meter is Hubbard's electro-psychometer to examine a person's mental state. It is similar to a crude polygraph. It is mainly used by a trained Scientology auditor—who is also trained well in hypnosis.[132]

Auditing: Auditing in Scientology claims to identify spiritual distress from a person's present and past lives. Auditing is a fake psychotherapy performed by a Scientology auditor. After putting the subject in command hypnosis, the auditor asks various questions to confuse the subject and injects some crazy ideas that the subject never went through. The E-meter measures the responses. This process continues from "pre-clear" to "clear" a couple of times until the subject is clear. And with each auditing, a significant sum of money goes to the Church of Scientology.

Thetan: Thetan is an immortal spiritual being, or human soul, or self. A thetan is a creator of things. *Thetan* is derived from the Greek letter Θ, theta. In Scientology belief, this means the source of life. Scientology believes each person is an immortal being, a force that believers call thetan. On a higher level, Scientology confronts body thetans through more auditing. Finally, the sacred teaching of Scientology or the Operating Thetan assists the individual in exercising as a fully conscious and functioning thetan.

Beliefs: The foundation of Scientology is based on a sequence of psychology and the manner the mind seems to work. The word *engram* is part of Scientology terminology. It is a memory trace or a hypothetical permanent change in the brain accounting for the existence of memory not available to the conscious mind, which was proved false by Karl Lashley. Scientologists erase their mind of engrams by taking part in therapy sessions called audits. For measuring engram, the auditor uses an E-meter. This process continues with a trained auditor until a person becomes "clear." A clear person is supposedly free of engrams or reactive minds and finally joins the thetan.

Scientology has the concept of God under the basic command, "Survive!" obeyed by all of life and expressed as the Eight Dynamics of Life,[133] which means the urge toward existence as Infinity; it is also called Supreme Being. The Eight Dynamics are the following:

1. Self
2. Creativity
3. Group Survival
4. Space
5. Life Forms
6. Physical Universe
7. Spiritual
8. Desire toward existence as Infinity

These eight dynamics were written in *Science of Survival* by L. Ron Hubbard.[134] Scientologists believe that this world emerged billions of years ago. It is populated with thetans who lived many lives before their present existence.

Worship and practices: The Church of Scientology ministers perform the same type of services and ceremonies as many other priests and ministers of other religions. Ministers must graduate from an authorized training curriculum. Scientology auditors must become ordained ministers. Scientology parishioners celebrate namings, weddings, christenings, and funerals. Scientologists do not have a personal deity. Therefore, they do not have private rituals of worship.

2—Pentecostalism

Pentecostals or Charismatics: Pentecostalism or charismatic movement is a worldwide movement that emerged under the leadership of Charles Fox Parham, an evangelical Protestant, in the early twentieth century. It is a kind of Christianity that emphasizes the personal experience of God through the Holy Spirit, called baptism with the Holy Spirit, and not through rituals; it is energetic and dynamic. They believe those baptized with the Holy Spirit can receive supernatural gifts to heal others, prophesy, speak in tongues, and gain many more blessings than other Christian denominations. Pentecostals do not have a single organization. The individual congregation came together to found various denominations. Pentecostalism expanded in the twentieth century around the world to over 250 million adherents; some suggest the number is more than 500 million.[135] They are not united in one single denomination in spite of believing in the baptism of the Holy Spirit, exorcism, speaking in tongues, faith healing, having supernatural experiences, and shared beliefs in selected doctrines of the Christian faiths. They all have strong beliefs in the literal interpretation of the Bible. Pentecostal and charismatic denominations include the following:

- The Church of Jesus Christ of Latter-Day Saints
- Apostolic Church
- Assemblies of God
- Church of God in Christ

- Church of God of Prophecy
- The United Pentecostal Church
- Pentecostal Holiness Church
- Fire-Baptized Holiness Church
- International Church of the Foursquare Gospel
- United Pentecostal Church International
- Vineyard Churches
- Full Gospel Baptist Church Fellowship

In 1893, Charles Fox Parham joined a Methodist Church in Kansas. He soon became enmeshed in holiness theology and thought he could do better by becoming an independent holiness evangelist and teacher. In 1898, he opened the Bethel Bible School and healing home in Topeka, Kansas. Under the direction of Parham, the students worked on speaking in tongues, or glossolalia. In 1901, a student called Agnes Ozman[136] spoke in tongues. It was a massive success for Parham. Sensational stories appeared in the local newspapers in Topeka and beyond.

By 1906, the Pentecostal movement had reached Los Angeles through one of Parham's students, William Seymour, a Holiness Church pastor. Under the leadership of Seymour, the Spiritual Center became famous and very wealthy with the performance of speaking in tongues.

In the beginning, Parham was part of the center, but soon he was expelled from the center by Seymour and an elder. In 1907, Parham was charged with sodomy in San Antonio during a local healing crusade. Parham retired to his home in Baxter Spring, Kansas, and died in 1929.

Christians believe that the Pentecost commemorates the descent of the Holy Spirit upon the apostles and other followers, and tongues of fire rested on their heads, in fulfillment of the promise of Jesus Christ (Acts 2:2–13). The manifestations of the Holy Spirit seen by the early believers include messages of wisdom, knowledge, faith, a gift of healing, miraculous powers, discerning of spirits, tongues, and interpretation of tongues (Act 2:4; 1 Cor. 12:4–10, 12:28).

Speaking in tongues is not a spiritual gift (Mark 16:17; Acts 2:1–12; 1 Cor. 12,14). It is not a beautiful work of God. Speaking in tongues is just learned behavior. It is a rapid and inarticulate talk with insignificant words, an unintelligible language. Many studies have been done,[137] and the students learn by a couple minutes of listening. This is one of the reasons the Pentecostal believers perform speaking in tongues more than any other church to influence the believers by showing emotional overtones.

The name "Holy Rollers" comes from practitioners who were rolling down the aisles of the church in their happiness and emotion. The believers in the church watched the practitioners speaking in tongues, with the accompanying excitement, and believed it was a way of communicating directly with God.

Festivals: Pentecost (Greek: *Pentekoste*, meaning fiftieth) is a Christian festival; it is a celebration of the day the Holy Spirit descended upon the disciples of Jesus.

It is also called Shavuot, a Jewish spring festival. This festival represents the birth of the early church and celebrates on the Sunday fifty days after Easter.

3—Hare Krishna

Hare Krishna's beliefs are based on the Hindu text Bhagavad Gita, or the "Song of the Lord," written around 250 BC. These songs were not only beloved to Gandhi but also all Hindus. Hare Krishna, also named Gaudiya Vaishnavism or Chaitanya Vaishnavism, is promoted by the International Society of Krishna Consciousness (ISKCON), and was first founded in the United States in 1965 by A. C. Bhaktivedanta. A. C. Bhaktivedanta, also called Swami Prabhupada, was born on September 1, 1896, in Calcutta, India, and died on November 14, 1977, in Vrindavan, Uttar Pradesh, India.

In 1922, he met a well-known spiritual master, Srila Bhaktisiddhanta Sarasvati Thakur, and became his sincere disciple. Srila Bhaktisiddhanta asked A. C. Bhaktivedanta to go to the USA to teach Krishna consciousness. In the meantime, he started a successful pharmaceutical business to support his family. Soon, he renounced all his family ties and gave his business to his son in 1954 to allocate all his life to his religion.

He followed his leader's advice for going to the land of opportunity. He was seventy years old when he arrived in New York City, and he rented a small storefront at 26 Second Avenue. His first converts were all hippies. They shaved their heads and adopted long Indian clothing. In 1967, A. C. Bhaktivedanta moved to the Haight Ashbury district of San Francisco, where he had tremendous success and gained many followers.

Hare Krishna is a mystical sect of Hinduism. Krishna was the eighth and principal avatar or incarnation of the Hindu god Vishnu, and was a supreme god because of his ability. Hare Krishna, or Bhakti, meaning a "devotion" or "participation" yoga movement, goes back to the sixteenth century. Bhakti yoga's founder, Chaitanya Mahaprabhu (1485–1534), advocated a devotional method of faith through recurrent dancing, chanting, with the saffron-colored robes, book-hawking, and especially of the Hare Krishna mantra:

Hare Krishna, Hare Krishna
Krishna Krishna, Hare Hare
Hare Rama, Hare Rama
Rama Rama, Hare Hare

Hare is the name of Srimati Radharani. It is calling him out to Sri Radharani, who steals the heart of Krishna by her infinite devotion and love. Krishna is depicted in many forms and means the all-attractive personality of the godhead.

A. C. Bhaktivedanta expected his Hare Krishna devotees to follow an ascetic life, not use alcohol and drugs, not eat meat, keep celibate, and save energy for the creation of children until marriage.

The mission of this Hare Krishna movement is to promote the welfare of society by teaching the disciplines of the Hare Krishna consciousness. It was issued by A. C. Bhaktivedanta in July 1965 in New York City. The following are the seven purposes of ISKCON:

1. To systematically propagate spiritual knowledge of society at large and to educate all people in the techniques of religious life to check the imbalance of merits in life and to achieve real unity and tranquility in the world.

Newer Religions

2. To propagate a consciousness of Krishna (god) as it is revealed in the great scripture of India, Bhagavad Gita, and Srimad Bhagavatam.
3. To bring the members of society together with one another and closer to Krishna, the prime entity, thus evolving the idea within the members and humanity at large that each soul is part and parcel of the quality of the godhead or Krishna.
4. To teach and encourage the Sankirtana (performing art singing, drumming, and dancing) movement, congregational chanting of the holy name of god, as revealed in the teachings of Lord Sri Caitanya Mahaprabhu (Bengali Hindu mystic and saint).
5. To erect for the members and society at large a holy place of transcendental pastimes dedicated to the personality of Krishna.
6. To bring the members closer together to teach a more straightforward, more natural way of life.
7. To achieve the purposes as mentioned above, to publish and distribute periodicals, magazines, books, and other writings.

Hare Krishna followers worship in and often reside in ashram temples.

Who Is Hare Krishna?

Krishna is a major Hindu god, and he was the eighth and principal avatar or incarnation of the Hindu god Vishnu. Krishna's adventures are narrated in sacred scriptures.

1. Mahabharata is one of the two ancient epic poems in human history. The other is the Ramayana. The poem is made up of almost 100,000 couplets and 1.8 million words. Its composition goes back to 900 BC; it was not in writing for a couple of centuries. The form of the poem is

The History of the Rise and Fall of the World's Religions

attributed to sage Vyasa and tells a story of the war between a family of brothers. The date and the occurrence of the war are much debated.
2. Bhagavad Gita is composed in the form of a conversation between Prince Arjuna and Krishna. It was written perhaps in the first or second century CE; it is commonly known as the Gita. They discuss the importance of selflessness, dharma (devotion), and the pursuit of moksha (liberation).

The primary source for the understanding of the historical meaning of Hinduism can be found in the sacred Vedas. The earliest holy texts, Vedas were possibly introduced between 1200 and 1000 BC to Indians by the Aryans who settled in northwest India around 1500 BC. Hindus believe that the texts were received by sages directly from God and passed on from generation to generation by word of mouth. The composition of these beautiful myths and stories was realized after at least one thousand years, and in these compositions are the myths of three most significant gods: Brahma, who creates the universe; Shiva, who destroys the universe; and Vishnu, who preserves the universe. They are called the Trinity or Trimurti.

The Krishna's story goes as follows: Mother Earth was tired of carrying the burden of the sins committed by evil kings and appealed to Brahma for help. Brahma asked Vishnu to annihilate the evil forces. Kasma, the king of Mathura, inspired fear among all the rulers. On the day Kasma's sister Devaki was married to Vasudeva, a voice from the sky forecasted that Devaki's eighth son would destroy Kasma. Kasma jailed the couple and the first seven infants of the couple. Lord Vishnu came to them and told them he would return to earth on the appearance of their son and rescue them from Kasma's cruelty.

As soon as the divine baby was born, Vasudeva found himself freed from prison. He fled with the baby to a safe house along the way. In the meantime, Vishnu cleared the way for Vasudeva. Vasudeva gave the baby Krishna to a cowherd's family, exchanging him for a newborn girl. Vasudeva returned to prison with a baby girl. When Kasma heard about the birth, he rushed to the prison to kill the baby girl. When he arrived, the baby girl ascended to the heavens and was transformed into the goddess Yogamaya. She notified Kamsa that his archenemy had already been born in another place.

Krishna was raised as a cowherd; as he matured, he became a good musician, chasing the women of his village and playing the flute. Finally, he returned to Mathura, where he killed his maternal uncle, King Kasma, and his cruel supporters. He restored his father to power, and became a good friend with Hinduism's heroes and warriors, like Arjuna.[138] He completed his education and mastered the sixty-four arts and sciences in sixty-four days. He stayed in Mathura until the age of twenty-eight.

Krishna lifted Mathura to rescue a clan of Yadava chiefs who were forced out by King Jarasandha of Magadha. He defeated Jarasandha and built a powerful capital, Dwarka, on an island in the sea. While his relatives and the natives slept, Krishna moved them to Dwarka by the power of his yoga and married Rukmini, Jambavati, and Satyabhama. He saved his kingdom from Narakasura, the evil king of Pragjyotisapura, who abducted sixteen thousand princesses. Krishna freed them and married them all. Krishna lived with the Pandava and Kaurava, who governed over Hastinapur.

A war broke out between the Pandava and Kauravas; Krishna tried to mediate and failed. Krishna offered his forces to the Kauravas, and he agreed to join the Pandavas as the charioteer of the master warrior Arjuna.

This epic Battle of Kurukshetra is explained in the Mahabharata, which was fought in around 3000 BC. During the war, Krishna conveyed his prominent advice, which is the core of the Bhagavad Gita, in which he put the rule of action without attachment or Nishkam Karma.

After the Battle of Kurukshetra, he returned to Dwarka. In his final day on earth, he taught spiritual wisdom to his best friend, Uddava, and other disciples, and ascended to his permanent abode after casting off his material body, which was shot by Jara, a hunter.

4—Rastafarianism

Rastafarianism arose in response to the harsh and inhuman treatment of slaves by the slave owners. The origins of Rastafarianism go back to the eighteenth century, when Pan-Africanism and Ethiopian movements highlighting an idealized Africa started to take place among Black slaves on the American continent. As a result of this movement, Marcus Garvey (1887–1904) preached that the African slaves were going back to Africa. He founded the Back to Africa movement, and preached that a future Black African king would lead the Africans.

In the meantime, the 1920s brought the Holy Piby (the Black Man's Bible), a proto-Rastafarian text, written by Robert Athlyi Rogers (May 6, 1891–August 24, 1931), who founded an Afro-centered religion in the United States.

The rise of Rastafarianism began when on November 2, 1930, Ras Tafari Makonnen was crowned King of Ethiopia; Ras Tafari means "chief dread," also called Rastafari. In the Ethiopian belief, Haile Selassie is part of the continuous line of succession of the biblical King Solomon and the Queen of Sheba, therefore connecting him to the tribes of Israel.

Soon, Ras Tafari claimed the title of Emperor Haile Selassie I, meaning "King of Kings, Lord of Lords, the conquering Lion of the Tribe of Judah, Elect of God, and Power of Trinity," which correlates to Revelation 19:16. Thus, to some, Haile Selassie fulfilled the biblical prophecy of a Black king that had been preached by Marcus Garvey in Jamaica. Also, Garvey thought that Ethiopia is the origin of the Black race. Ethiopia was so important to him because Haile Selassie was the only independent Black ruler at the time on the African continent.

The religious movement Rastafarian takes its name from Emperor Haile Selassie's pre-coronation name, Ras Tafari. During the reign of Haile Selassie, preachers such as Joseph Hebbert, Robert Hinds, Archibald Dunkley, and Leonard Howell achieved prominence. Among them, Leonard Howell was called the father of the Rastafarian movement. He was later arrested a few times by the Jamaican authorities for preaching revolutionary doctrine. In 1933, Leonard Howell supplied the Rastafarians the following six principles:

1. Hatred for the White race
2. The complete superiority of the Black race
3. Revenge on Whites for their wickedness

4. The negation, persecution, and mortification of the government and legal bodies of Jamaica
5. Preparation to go back to Africa
6. Acknowledging Emperor Haile Selassie as the Divine Being and only potentate of Black people

Also, he noted the following commitments for Rastafarian believers:

1. If one claims to be Rasta, then should they not follow the principles of the father of the Rastafarian movement, Leonard Howell?
2. Could it be due to the fact of White Rastas gaining momentum in our movement that we have found ways to compromise with Babylon?
3. No compromise with I and I! I and I means that all are equal.
4. I follow the teaching of Leonard Howell.

In the late 1940s, another version of Rastafarianism, called the Youth Black Faith, emerged in the capital of Jamaica, Kingston, to the existing Nyabinghi Mansion or branch. This branch was aggressive toward the Jamaican authority. This branch introduced some new features to become part of Rastafarians, such as growing their hair into dreadlocks and adopting the group's unique dialect.

Emperor Haile Selassie reportedly refused the Rastafarian painting of him as a god. Soon after the refusal of being their god, in 1948, he embraced their cause and donated five hundred acres named Shashamane, meaning "gift of God" for those Black slaves who came back to their homeland, Ethiopia.

In 1958, Prince Emanuel Charles Edwards founded the Ethiopian International Congress, or Bobo Ashanti.

In 1968, Vernon Carrington, a.k.a. the Prophet Gad, founded the Twelve Tribes of Israel, and encouraged Blacks to return to their homeland.

From 1960 to 1970, Rastafarians gained respect in Jamaica and visibility in other parts of the world through the popularity of

Rastafarian-inspired reggae musicians like Bob Marley, Jimmy Cliff, Jamaica Etana, Takana Zion, and more.

On April 21, 1966, Emperor Haile Selassie I visited Jamaica for the first time. Around one hundred thousand Rastafarians from every part of Jamaica descended on Palisadoes Airport in Kingston, Jamaica, to see the man they considered to be their Jah or God, who was arriving to visit them. On his visit, two critical things happened:

1. April 21 was declared a holy day called Grounation Day.
2. Haile Selassie encouraged Rastafarians through Mortimer Planno and Joseph Hibbert not to immigrate to Ethiopia until they had first liberated the people of Jamaica.

Rastafarianism endured a considerable setback at the result of Haile Selassie's death in 1975 and Bob Marley's death in 1981. The loss of these two most influential figures affected the progress of Rastafarianism in the United States, the United Kingdom, Africa, and the Caribbean. There are an estimated one million Rastafarians around the world, with the majority living in Jamaica.[139]

Rastafarianism Beliefs

Rastafarianism arose in response to the harsh and inhuman treatment of slaves by slave owners, where the Christianity and Islam interpreted the Bible and Koran to benefit the slave owners. It was a social and political movement, and mostly the culture of the world was such that a movement like this was called religion. Rastafarians do not have a formal creed, but the preacher Leonard Howell offered six principles of Rastafari, which include potent expressions of hatred, revenge, and nostalgia back to Africa.

Rastafarians accept most of the Bible but with some reservations. They believe that much of the translation into English has been corrupted. In the meantime, the basic text may be right, but it should be inspected in a particular light. Mainly, they accept the prophecies in the New Testament and the book of Revelation regarding the Second Coming of the Messiah. They believe the Second

Coming has already happened in the form of Haile Selassie. As the incarnation of God or Jah, Haile Selassie is both king and God to Rastafarians. While Haile Selassie officially died in 1975, many Rastafarians do not believe that God or Jah can die, and therefore, they think it was fake news. Others believe that he is still alive in the spirit but not in a physical form.

Rastafarians are deeply influenced by Judeo-Christianity and share many commonalities with both Judaism and Christianity. They believe the Black race is one of the Israelite tribes, and they are part of the chosen people. They follow many Old Testament rulings, such as the forbiddance of cutting one's hair. That is why they use the dreadlocks as their symbol for not cutting hair. Also, they do not eat pork or shellfish. Some believe that the Ark of the Covenant is placed somewhere in Ethiopia. The term *Babylon* is pulled from biblical stories of the Babylonian captivity of the Jews, and is associated with an unjust and oppressive society. Rastafarians use this term about White society, which enslaved Africans and their descendants for centuries.

In addition to dreadlocks, there is another symbol for Rastafarians, and that is ganja, which is a strain of marijuana. They view ganja as a spiritual purifier. They smoke ganja to cleanse the body and open the mind; this is based on different verses from the Bible:

1. "He causeth the grass for the cattle and herb for the service of the man: that he may bring forth food out of the earth" (Ps. 104:14).
2. "Thorns also and thistles shall it bring forth to thee, and thou shalt eat the herb of the field" (Gen. 3:18).
3. "Better is a dinner of herbs where love is than a stalled ox and hatred in addition to that" (Prov. 15:17).

The use of ganja is not only for spiritual purposes, but it is also used as medicine for colds. Many Rastafarians limit their diets to what they believe to be pure food. They do not use certain additives, alcohol, or coffee, other than ganja.

Rastafarians have holy days and celebrations.

1. January 7, Orthodox Christmas
2. April 21, Grounation Day
3. July 16, Ethiopian Constitution Day
4. July 23, Emperor Haile Selassie's birthday
5. August 17, Marcus Garvey's birthday
6. September 11, Ethiopian New Year's Day
7. November 2, Coronation Day of Emperor Haile Selassie

5—Shamanism

Shamanism originated in northern Asia, and it is generally agreed that the origination goes back to the hunter-gatherer cultures; it is included in this section of newer religions due to its popular resurgence in some cultures (like Native American). It persisted within some herding and farming communities after the Neolithic period or agricultural revolution to the present time. Through the history of Shamanism, members of tribes would gather and experiment with the poisonous and very highly psychoactive mushrooms, like the fly agaric or fly amanita or amanita muscaria. Once these members used these mushrooms to learn the effect, they were classified as shaman. There is no doubt that many cultures around the world may have learned from previous tribes or conducted similar practices.

Shamanism is a form of animistic religion that comprises a belief in the power of a shaman. Over the past few decades, with the popularity of the New Age beliefs, Shamanism captured a lot of attraction and growth in the Western world and, surprisingly, in the United States mostly due to the Native American tribes that still practice Shamanism.[140]

A shaman is a person who is believed to possess special skills to influence spirits. A shaman can be a man, a woman, or a transgender person of any age from middle childhood onward who can be either warmhearted or injurious. Shamans are known for their contact with a supernatural phenomenon, such as the world of gods, demons, and ancestral spirits. They gained a reputation as healers and foretellers of the future with good spirits, while some cause diseases, including mental disorders, in the tribal community. Most shamans are called medicine men or witch doctors, with respect.

The word *shamanism* is said to come from the Manchu-Tungus language word *šaman*. The noun is created from the verb *ša*, meaning "to know"; therefore, a shaman is precisely "one who knows." Some experts believe that the term *shamanism* may apply to all religious systems in which a central bigwig is trusted to have direct communication with the supernatural being that allows that central bigwig to act as a diviner and the like. These types of reactions can happen only through a trance state, hallucination, and ecstasy. These actions are psychosomatic phenomena that can be brought about at any time by the person who is trained with the ability to act, and that is a shaman.

Shamanism Belief

The supernatural is the kingdom of shamans. They believe they are linked to the spirits to heal, uplift consciousness, contact deceased ancestors, and influence the weather. Scholars agree that the shaman may either inherit the profession or have an inner calling. Regardless of the type of inheritance or inner calling, this profession requires some qualification.

Once the decision has been made, the future shaman has to launch into a period of intensive training. They have to learn some specific practices under the supervision of active shamans, including the techniques of the trance state through autohypnosis, fasting, and the ingestion of hallucinogens. A shaman's professional proficiency can be shown by the personal abilities, including dramatic talent, mental capacities, and power to make their will effective.

There are many dissimilarities of Shamanism around the world within different cultures. Based on Mircea Eliade's research of 1972,[141] some common beliefs by all divergences of Shamanism have been found.

1. Spirits exist, and they play essential roles both in individual lives and in human society.
2. The shaman can share information with the spirit world.

3. Spirits can be benevolent or malevolent.
4. The shaman can cure sickness caused by evil spirits.
5. The shaman can employ trances and persuasion techniques to stir up visionary ecstasy and go on vision quests.
6. The shaman's spirit can leave the body to enter the supernatural world to search for answers.
7. The shaman evokes animal images as spirit guides, omens, and message-bearers.
8. The shaman can perform other varied forms of divination, such as scrying (crystal gazing),[142] throwing bones or runes, and sometimes foretelling future events.

Shamans are believed to visibly permeate the supernatural, which affect or influence the lives of all living things. Some malicious spirits cause diseases to humans, and shamans use physical and spiritual methods to heal the patients. Most of the time, the shaman enters the body of the patient to confront the malicious spirit and treats by expelling the evil spirit.

Shamans are very important at three significant life events: birth, marriage, and death.

If a woman has not bore a child, the shaman goes to heaven and sends the woman an embryo soul from the tree of embryos. Also, the shaman executes oblation after birth so the child grows faster and doesn't cry.

The spirit spouse is one essential ingredient of Shamanism. The shaman assists the spirit husband and wife to work, and they gain strength in the world of the spirit.

When somebody dies in the community, the shaman is needed to catch the soul of the deceased wandering in the universe and guide it to the Remote World.

Shamans wear a particular dress to imitate an animal, like a bird, deer, or a bear, with a headdress made of antlers and feathers of birds. Dance, song, and drama were the rituals of hunter-gatherers and continue to be part of many religions and cultures. In Shamanism, a shaman is, at the same time, a dancer, singer, and actor. The shaman's songs contain many obligatory conversations, images, refrains, and smiles to influence the patient.

Summary

The newer religions come from the established religions of Christianity, Hinduism, or the primitive religions.

Leadership-followership evolved to allow and assist the species in sharing information and coordinated group behavior for their survival. Among a group, one individual who has the psychological, physical, behavioral, and social capital will emerge as a leader, and the rest will be followers. The leader could be good or evil.

1. *Scientology*: Scientology International teaches that people are immortal spirits who have forgotten their true nature. Scientology is a philosophical/religious movement. L. Ron Hubbard founded Scientology in 1954 after failing to graduate from George Washington University in Washington, DC. He had the talent of a leader. L. Ron Hubbard created a corporation called Dianetics Foundation in 1950. He published his bestselling book *Dianetics: The Modern Science of Mental Health* through the Dianetics Foundation. Soon, the Dianetics Foundation went bankrupt. In 1954, he renamed the Dianetics Foundation to the Church of Scientology, and fooled his followers. This secretive church and only two of its many offshoot organizations are worth more than $1.2 billion.
2. *Pentecostals or charismatics*: The Pentecostal or charismatic movement isn't a religion on its own; rather they attach to some churches and influence the attendees by preaching the so-called prosperity gospel in some cases. It is a worldwide movement that emerged under the leadership of Charles Fox Parham, an evangelical Protestant, in the early twentieth century. They believe those baptized with the Holy Spirit can receive supernatural gifts to heal others, prophesy, and speak in tongues. Parham was charged with sodomy in San Antonio during a local healing crusade. Pentecostals have twelve denominations and five hundred million followers.
3. *Hare Krishna*: Hare Krishna was first founded in the United States in 1965 by A. C. Bhaktivedanta. Srila

Bhaktisiddhanta asked A. C. Bhaktivedanta to go to the USA to teach Krishna consciousness. A. C. Bhaktivedanta had massive success among hippies. He introduced Krishna, a major Hindu god, to his followers. Krishna was the eighth and principal avatar or incarnation of the Hindu god Vishnu. He achieved his goal in the land of opportunity and published and distributed periodicals, magazines, books, and other writings.

4. Rastafarianism: Rastafarianism arose in response to the harsh and inhuman treatment of slaves by slave owners. Marcus Garvey founded the Back to Africa movement, and preached that a future Black African king would lead the Africans. Rastafarians believe Christianity and Islam interpreted the Bible and Koran to benefit the slave owners. They believe the Second Coming has already happened in the form of Haile Selassie. As the incarnation of God or Jah, Haile Selassie is both king and God to Rastafarians.

5. Shamanism: Shamanism is one of the primitive religions that has regained popularity, and is a form of an animistic branch of primitive religion that comprises a belief in the power of a shaman. Scholars agree that the shaman may inherit the profession or answer an inner calling. A shaman's professional proficiency can be shown by their personal abilities, including dramatic talent, mental capacities, and power to make their will effective. Most shamans are called medicine men or witch doctors, with respect. In Shamanism, a shaman is, at the same time, a dancer, singer, and actor. The shaman's songs contain many obligatory conversations, images, refrains, and smiles to influence the patient. They believe in direct communication with the supernatural being; it happens only through trance state, hallucination, and ecstasy. Over the past few decades, with the popularity of New Age beliefs, Shamanism has captured a lot of attraction and growth in the Western world and, surprisingly, in the United States, mostly due to the Native American tribes that still practice Shamanism.

Credit: Coolvectormaker

CHAPTER 17

Atheism

Power of mind, what humans own (Copyright iStock by Getty Images/credit: Kateryna Kovarzh).

Atheism is derived from the Greek word *atheos*. A- is a Greek prefix meaning not or without, and *theos*, meaning God or deity, so as one word, it means godless or without God. Atheism, in general, is active disbelief in the existence of a god or gods.

Nonreligious people include agnostics, atheists, humanists, freethinkers, and seculars, and are the world's third largest population, with close to 1.2 billion,[143] after Christianity (2.3 billion) and Islam (1.8 billion).

Atheism historically goes back to the rise of Jainism and Buddhism, in the sixth century BC. Indians were tired of the caste system. The castes, in the order of prominence, include Brahmin, or the intellectual and spiritual leaders; Kshatriyas, or the protectors and public servants of society; Vaisyas, or the skillful producers; and Shudras, or the unskilled laborers. Buddha himself was from the Kshatriya class.

Jains and Buddha did not believe in the existence of God or gods. In China, Confucius, in the sixth and fifth century BC, did not believe in the presence of God or gods; he was an atheist with the highest morals. Lao Tsu's talks were about nature. He never mentioned the supernatural in his books, yet he did not deny them either; he was very well fit as an agnostic.

A Greek philosopher, poet, theologian, and social and religious critic, Xenophanes of Colophon (sixth century BC) is taken by some to be a pantheist. In contrast, others maintain that he was fundamentally an atheist or materialist.

In the fifth century BC, atheist Heraclitus of Ephesus said, "There is nothing permanent except change… This world order is the same for all, has not been made by any god or man. It always was and is and will be an ever-living fire, kindling by itself or by measure and going out by measure."

Later, during the fourth century BC, many Greek philosophers were profoundly thinking of the natural world and its operation, and they certainly believed that the deities of Olympus were meaningless. The uneducated public and believers of the Olympus deities punished those who were godless.

Greek philosophers of the fourth century BC, Epicurus, Leucippus, and Democritus, were the founders of the atomic theory. They expressed and believed that the universe was formed from natural origin and atoms colliding by accident, not by any creators. The Epicurean school of thought was criticized and insulted by the Abrahamic religions. Epicurus disbelieved the existence of God because of the presence of suffering and evils. The following Epicurus quote shows his disbelief in God. "Is God willing to stop evil, but not able? Then he is not omnipotent. Is he able but not willing? Then he is malevolent. Is he both able and willing? Then whence cometh evil? Is he neither able nor willing? Then why call him God?"[144] Atheism never faded with the rise of Christianity and Islam. Christianity and Islam both tried to impose the superiority of their belief by the power of the sword and killing, not reason, until the eighteenth century.

Atheism emerged before Enlightenment from the widespread rejection of myths and progress in science, which became a tool to the world based on arguments and evidence that are accessible to all. During the Reformation and Counter-Reformation, European states were fighting against the intolerance of religious minorities. Many lost their lives under the accusation of heresy. The unity of religious minorities and the majority was essential for social and political stability. The rise of atheism was the result of religious violence by the Catholic Church, like the Spanish Inquisition, the expulsion of the Huguenots from France, the civil wars of England, the witch trials, and civil wars in the Netherlands and Scotland. Many writers on tolerance emerged, like John Locke, an English philosopher and physician; Voltaire (Francois-Marie Arouet), a French writer and public activist; Benedict de Spinoza (Baruch Spinoza), a Dutch Jewish philosopher of Portuguese Sephardi origin; Pierre Bayle, a French philosopher and writer; Baron D'Holbach (Paul-Henri Thiry), a French-German philosopher who was famous for his deterministic and materialistic metaphysics; David Hume, a Scottish historian, philosopher, economist, and essayist; Denis Diderot, a French writer, philosopher, and art critic; Jean Le Rond d'Alembert, a French mathematician, physicist, philosopher, music theorist, and the editor of the famous Encyclopedie. They all encouraged the public toward religious tolerance. They argued for individual freedom to express their ideas and beliefs.

This progress, revolution, and Enlightenment happened only in Europe and North America. Islamic countries in Africa and Asia continue with intolerant behavior to this day.

Twentieth-century atheist Bertrand Arthur William Russel, third Earl Russell, OM FRS (1872–1970), was a British philosopher, mathematician, logician, writer, historian, social critic, political activist, essayist, and Nobel laureate. Russel was a polymath in that he grasped several fields of philosophy and the application of them. Russel's views, proposed in his book *Why I Am Not a Christian*, leads to an essential distinction between theology and philosophy. His book stimulated an intense backlash of intellectuals and pious readers. Russel reacted against the first-cause argument and the design and natural law arguments among the five arguments to prove the existence of God by Saint Thomas Aquinas, and Richard Dawkins proved all Five Arguments are wrong.

Twenty-first-century atheists, or the New Atheists, include Richard Dawkins, Sam Harris, Daniel Dennett, and Christopher Hitchens, who are authors of many books to promote atheism. The term "New Atheist" or "atheist of the twenty-first century" is a label for these authors and critics of religion and religious belief coined by the journalist Gary Wolf to explain the positions promoted by these authors. There is no significant difference in the atheism of Bertrand Arthur William Russel and those proposed by the New Atheism.

Types of Atheism

Atheists do not have a God, gods, or goddesses, beliefs, scripture, festivals, holidays, rituals, costumes, denominations, prophet, or leaders. Atheism is a straightforward concept of no God. They meet for social events in any building, and they have presidents or directors. There are several atheism types with some minor differences under a big umbrella of atheism.

New Atheism: Atheism, in general, is the absence of belief in the existence of God, gods, or goddesses. At the forefront of New Atheism are several talented writers, thinkers, scientists, and philosophers like Richard Dawkins, Daniel Dennett, Sam Harris, and

Christopher Hitchens. They advocate their view that belief in God and religion should not be tolerated but should be sharply criticized and contradicted with a rational argument when they try to influence the government, politics, and education. The New Atheists, metaphysically, share the belief that there is no divine reality of any kind. Epistemologically, they all agree that religious belief is irrational, and also morally, they agree that there is a universal secular moral standard. These values are in a December 7, 2005, manifesto published by Sam Harris.

The New Atheism advocates use the natural sciences in their criticisms of theistical belief and the same way in their proposed explanations of its origin and evolution.

Through the years, the Ten Commandments have been removed from a public place by order of the court. The New Atheists used the natural sciences in the battle over whether to teach creationism, now called intelligent design, in the schools or whether to teach Charles Darwin's theory of evolution.

This battle was fought in Dayton, Tennessee, in the Scopes Monkey Trial in July 1925. Scopes lost the case, and the fundamentalists pushed further; as a result, some states passed laws banning the teaching of evolution. The fight continued in 1968 in *Epperson v. Arkansas* and in 1987 in *Edwards v. Aguillard*.

In October 2004, the creationists framed creationism as a science with a new name, intelligent design, against the Dover Area School District of York County, Pennsylvania, which can teach it as an alternative to evolution theory. This case is sometimes referred to as the Dover Panda Trial. This case was tried in a bench trial from September 26, 2005, with a pretentious fight, to November 4, 2005, before Judge John E. Jones III, a Republican appointed in 2002 by George W. Bush. On December 20, 2005, Jones issued his decision, ruling that the Dover mandate was unconstitutional. The ruling declared that intelligent design is *not* science, and banned the board from teaching intelligent design within the Dover Area School District.

Furthermore, New Atheism continues to criticize irrationalities sharply. Richard Dawkins claimed without hesitation that the five ways or five proofs declared by St. Thomas Aquinas, a Catholic philosopher of the thirteenth century for the existence of God, do not prove anything. The first three arguments: the Unmoved Mover, the Uncaused Cause, and the Cosmological Argument, say the same thing with different ways and wordings. All three cases necessitate an infinite regress.

- Aquinas's first argument is the Unmoved Mover; he says that nothing moves without a prime mover. His reasoning leads us to an infinite regress, from which the only escape is God. A mover had to make the first move, and that mover we call God.
- Aquinas's second argument is the Uncaused Cause; he says that nothing is caused by itself. While every effect has an antecedent cause, this series of reasoning goes to an infinite regress. This infinite cause has to be terminated by a first cause, and we call it God.
- Aquinas's third argument is the Cosmological Argument; he says there must have been a time when no real thing existed. Still, it is clear just by looking around that material things exist now. Therefore, a nonphysical thing must have brought them into existence, and this we call God.
- Dawkins's response to this argument says, "All three of these arguments rely upon the idea of a regress and to call on God to terminate it. They make the entirely unwarranted assumption that God himself is immune to the regress. Even if we allow the dubious luxury of arbitrarily conjuring up a terminator to an infinite regress and giving it a name, simply because we need one, there is absolutely no reason to endow that terminator with any of the properties normally ascribed to God: Omnipotence, omniscience, goodness, creativity of design, to say nothing of such human attributes as listening to prayers, forgiving sins and reading innermost thoughts. Incidentally, it has not escaped the notice of logicians that omniscience and omnipotence are naturally incompatible. If God is omniscient, he must already know he is going to intervene to change the course

of history using his omnipotence. But that means he can not change his mind about intervention, which means he is not omnipotent."[145] Aquinas's fourth argument is that from the degree of perfection, we know that things in the world differ in their maturity, some greater or lesser. A human can be good or evil, so the maximum cannot rest on humans. Therefore, there must be some other supreme perfection that all other beings fall short of. In Aquinas's fourth argument, we call that maximum God.

Dawkins's response to this fourth argument says, "That is an argument? You as well say people vary in smelliness. Still, we can make the comparison only by reference to an absolute maximum of conceivable smelliness. Therefore there must exist a peerless stinker, and we call him God. Or exchange any dimension of comparison you like and derive an equivalently fatuous conclusion" (2008).

- Aquinas's fifth argument to prove God's existence is the Argument from Design. All inanimate and animate things in the world look as though they are designed. Nothing that we know looks designed unless it is intended. Aquinas's fifth argument is there must be a designer, and we call him God.

Dawkins's response to the fifth argument says, "The argument from design is the only one still in regular use today, and it still sounds to many like the ultimate knockdown argument. The young Darwin was impressed by it when, as a Cambridge undergraduate, he read it in William Paley's Natural theology. Unfortunately for Paley, the mature Darwin blew it out of the water. There has probably never been a more devastating rout of popular belief by clever reasoning than Charles Darwin's destruction of the argument from design. It was so unexpected. Thanks to Darwin, it is no longer true to say that nothing that we know looks designed unless it designed. Evolution by natural selection creates an excellent simulacrum of design, mounting prodigious heights of complexity and elegance" (2008).

Humanism: Humanism searches to foster broad well-being by advancing self-determination, equality, and compassion to allow individuals to thrive and be able to live in a society with one another. These values are driven from human experiences and are available in two manifestoes published in 1933 and 1973[146] and defined as follows: Humanism is a rational philosophy of life that, informed by science. It rejects the existence of God and other supernatural beliefs. It stresses an individual's dignity and worth and capacity for self-realization through reason. Humanism states our responsibility and ability to lead ethical lives of personal achievement that aspire to the greater good.[147]

Agnosticism: Agnosticism comes from the Greek compound words *a-* meaning non and *-gnosis*, meaning knowledge. Agnostic means "I do not know." Thomas Henry Huxley used the term for the first time in the nineteenth century. He defined it: "Agnosticism is not a creed but a method, the essence of which lies in the vigorous implementation of a single principle. Positively the principle may be expressed as in matters of intellect; do not pretend conclusions are certain that not demonstrated or demonstrable."[148]

There are two types of agnostics—one a weak agnostic, and the other a robust or strong agnostic. The weak agnostics have no firm beliefs about God. They are unphilosophical and noncommittal. It can be personal and confessional, as with "I do not know if there is a God; I might find out in the future." The strong agnostics are philosophical committals. They view God's existence as unknowable, and no one can make a negative or positive defense for the existence of God.

Freethinker: This term was first used in England by Irish natural philosopher William Molyneux (1656–1698) in a letter to John Locke a year before his death. In that letter, he called John Toland (1670–1722) "a candid freethinker" who opposed the literal belief in the Bible and the Church.

Fifteen years later, the term appeared in print again from the Church of England and referred to "Atheists, Libertines despisers of religion under the name of freethinkers."[149] Very famous philosophers like John Locke and Voltaire were called freethinkers, and a magazine, *The Freethinker*, has been published in England from 1881 to the present. According to the Cambridge Dictionary, a freethinker is "someone who forms their own opinions and beliefs, especially about religion or politics, rather than just accepting what is officially or commonly believed and taught." Freethinkers are naturalistic and strongly reject religious beliefs.

Morality

Natural selection equipped all living species with hardwired emotions like fear, happiness, frustration, anger, hunger, love, moral sense, and sexual desire for their survival, and took millions of years for improvement. Among all the living species, human beings developed much better. They are social, and human sociality is the result of evolution, not a choice.

First, naturally, humans understood and knew if a person hurt somebody, that person will see a reaction and would get destroyed by revenge. This revenge brought the idea of morality to the very primitive society, possibly during the age of Homo habilis, who lived on the earth 2.5 million years ago.

Second, this goes back to the age of Homo sapiens, about 117,000 years to the present time, which was the first human society called hunter-gatherer. The hunter-gatherer faced natural disasters like floods, earthquakes, and storms, which caused a lot of suffering to them, like killing their loved ones and destroying their food sources. Repeated calamities made them think and search for protection from ancestors, natural objects like the sun, and another species, like a snake. As a result, worship started to come into people's minds.

As is evident from the above statements, first, morals entered human life from the time humans walked on the earth, and later religion was created from the fear of natural disaster for survival. It concludes that there is no relation between religion and morality. Morality differs based on culture and time. As earlier described, humans are social beings for their survival; they need to build groups to improve cooperation. These groups use particular codes of morality for welfare and interaction. Along with those codes of morality come some new rituals, like dancing after successful hunting; this is the beginning of religion. From this point, religion and morality go shoulder to shoulder. For a review, the universal moral codes of various religions independent of the existence of God, which is called the Golden Rule, or doing unto others as you would have them do unto you, please see the section "Morality" in chapter 12.

It is clear from observation of the Golden Rule of various religions that the moral codes working to minimize the suffering of others arises in all faiths. Among all religions, Jainism, which does not believe in God, ranks the highest in morality.

In conclusion, morality and religion are two separate things. Morality is hardwired in the human brain and makes the distinction between right and wrong, and is an evolved faculty with a genetic basis.

Summary

Atheism is derived from the Greek word *atheos*. *A-* is a Greek prefix meaning not or without, and *theos* means God or deity, so as one word it means godless or without God. Atheism, in general, is active disbelief in the existence of a god or gods.

Atheism historically goes back to the rise of Jainism and Buddhism, sixth century BC. Jains, Buddha, and Confucius did not believe in the existence of God or gods. They were atheists with the highest morals. Lao Tsu's talks were about nature. He never mentioned the supernatural in his books.

Christianity and Islam both tried to impose their superiority of their belief by the power of the sword and killing, not reason, until the eighteenth century. Atheism never faded with the rise of Christianity and Islam.

CHAPTER 18

Ecumenism

Ecumenist Christian missionaries converted East Asians who were born atheist to Christianity (Copyright iStock by Getty Images/credit: Doldam10 and Aurorat).

Ecumenism, also called ecumenical movement or ecumenical tendency, aims to promote worldwide Christian unity or cooperation. Ecumenism is derived from Greek word *oikoumene*, which means the whole inhabited world, and can be traced to the New Testament meaning the Roman Empire (Luke 2:1), the world as a whole (Matt. 24:14), the kingdom of the world (Luke 4:5), or the world bound to be redeemed by Jesus (Heb. 2:5). The perception of one Church serving God in the world is a reflection of a central teaching of the early Christian belief.

The early Church had many conflicts that called for ecumenical declaration. Continuing conflicts among Christians started in the second century. First, the Gnostics proposed several doctrines, among them that God did not create the world, but instead his divine did and moved out of the Church. Second, the Quartodecimans disputed the date of the Easter festival, which Rome called for on a Sunday and the Asia Minor Church called for on fourteenth Nisan. Later, another argument developed between two interpretations of Homousians and Arians, or Trinitarian and non-Trinitarian.

There have been seven ecumenical councils to resolve conflicts.

- *The First Council of Nicaea*: In 325 CE, the first ecumenical council of the Christian Church was called by Emperor Constantine I in Nicaea, now Iznik, Turkey, who hoped the council could solve the problem created by Arianism. This problem continued until the Second Ecumenical Council in 381. The first real version of the Bible was approved by the First Council of Nicaea in 325 CE. Athanasius, the bishop of Alexandria, wrote the approved Bible in 367 CE. Prior to this date, there was no written Bible.
- *The First Council of Constantinople*: In 381 CE, the Second Ecumenical Council of the Christian Church was called by Emperor Theodosius I in Constantinople, Turkey. The Council of Constantinople declared the Trinitarian doctrine of the equality of the Holy Spirit with the Father and the Son, and affirmed the Council of Nicaea.

- *The Council of Ephesus*: In 431, the Third Ecumenical Council of the Christian Church was called by Emperor Theodosius II in Ephesus, near Selcuk, Turkey. The Third Ecumenical Council approved the original Nicene Creed, which is the enlarged version of the Creed of Nicaea, and condemned the teaching of Nestorius, patriarch of Constantinople, who said the Virgin Mary might be called the birth giver of Christ, but not the birth giver of God. Nestorius was condemned for heresy and exiled to Upper Egypt.
- *The Council of Chalcedon*: In 451 CE, the Fourth Ecumenical Council of the Christian Church was called by Emperor Marcian and was held in Chalcedon, Turkey. About 520 bishops or their representatives attended the council. This council was the best documented and the largest of past councils. It affirmed the Creed of Nicaea in 325 CE, the Creed of Constantinople in 381 CE (known as the Nicene Creed), two letters of Saint Cyril of Alexandria in opposition to Nestorius, and Tome of Pope Leo I. This council described these doctrines in its confession of belief.
- *The Second Council of Constantinople*: In 553 CE, the Fifth Ecumenical Council of the Christian Church was called by Eutychius, patriarch of Constantinople. Pope Vigilius of Rome did not attend. He later ratified the verdict of the council on February 23, 554 CE. The fourteen anathemas provided by the council rejected Nestorianism and also confirmed an earlier condemnation of Origen, a Christian writer, teacher, and mystic.[150]
- *The Third Council of Constantinople*: In 680 CE–681 CE, the Sixth Ecumenical Council of the Christian Church was called by Emperor Constantine IV for a meeting in Constantinople. This council was composed of 170 bishops. The council condemned the false doctrine of heretics, Monothelites, including Pope Honorius I, and asserted two operations and two wills of Christ. The Eastern Christians were mostly Monothelites who were forbidden to talk of Monothelitism.

- *The Second Council of Nicaea*: In 787, the Seventh Ecumenical Council of the Christian Church was called by Empress Irene, widow of Emperor IV, in Nicaea, Turkey. This council was composed of 367 bishops. This council convened against the iconoclastic controversy heresy that arose in 726 by Byzantine Emperor Leo IV, who wanted to convert the Muslims to Christianity. He proclaimed the veneration of icons. The council declared decorum of images of Jesus Christ, the Virgin Mary, the holy angels, other pious and holy men, and also deposed all the bishops and clergy who refused to allow icons in the Church. The iconoclastic controversy continued until 843 CE when Empress Theodora came to power and restored the use of symbols following the Second Council of Nicaea. It was the last council to be recognized by both the Catholic and Eastern Orthodox Churches.

The iconoclastic controversy continued between the Byzantine Empire and the Roman Empire until complete separation in 1054 CE. The division of Christianity created a nightmare in the Christian world. It caused warfare and bloodshed. Some say it's hypocritical for Christians to look down on the conflict between Shiites and Sunnis in the Muslim religion, which killed thousands, while Christians killed thousands too.

In the meantime, from 630 CE, the Islamic empire invaded part of the Byzantine Empire territories, and thousands of Arabs and Christians were killed. The Arabs reached France, and the Roman Empire had no choice other than to attack the Muslims. The first Crusade began in 1095 CE and the final Crusade ended in 1271 CE, with tremendous losses on both sides during 200 years of war.

As noted above, for over 1,100 years, many ecumenical councils of the Christian Church were convened. Finally, the Church separated into two churches in 1054 CE, Eastern Orthodox Church and the Roman Catholic Church.

The idea of ecumenical tendency aimed to do good deeds, such as give help to the impoverished and victims of war, fight oppression, and aid in recovery from natural disasters. On the other side, politically, the thinkers of the ecumenical movement wanted to consolidate power and get rich. The religious institutions became so strong that the Roman Catholic Church controlled the entire European continent. The pope was installing and removing heads of state for centuries, and the public was fed up and attempted to reform the Roman Church in 1517 CE.

Finally, the Roman Catholic Church split into the Catholic Church and Protestant Church. Still, ecumenism has deep roots among all Christian religions, especially Catholics. As a result, the modern ecumenical movement stepped in.

It is not clear where the modern ecumenical movement started. Some assume the 1910 World Missionary Conference in Edinburgh, Scotland, was the beginning of the ecumenical movement, and some believe that the Roman Catholic Church tried to reconcile with the Reformation churches that separated over theological differences. In this conference, there were about 1,400 participants, and there was no Catholic or Orthodox representative present.

William Temple (1881–1944), who was the archbishop of Canterbury in the Church of England, proclaimed, "The ecumenical movement is the great new fact of our era."[151] At the 1910 World Missionary Conference, eight commissions were presented.

1. Conveying the Gospel to all the non-Christian world
2. The Church in the mission field
3. Education concerning Christianization of national life
4. Missionary message concerning the non-Christian world
5. The preparation of missionaries
6. The home base of missions
7. Missions and government
8. Cooperation and the promotion of unity

The 1910 Conference led to many new ideas and gave birth to the following:

- The creation of the International Missionary Council in 1921.
- The Life and Work Movement in 1925 in Stockholm, Sweden. This conference was to provide aid for the victims

- of war, poverty, natural disasters, oppression, economic and social injustice, racism, and sexism.
- The First World Conference for Church Unity was part of the Faith and Order Movement. The conference was held in Lausanne, Switzerland, in 1927.
- This Ecumenical Movement Conference was in response to the communist revolution of 1917, and was held in Jerusalem. This conference discussed the relation between the Christian message and other faiths and, more importantly, the theological interpretation of Christian political and social involvement.

Many other attempts of the Ecumenical movement conference took place: in Tambaram, near Madra, India, in 1938; in Whitby, Canada, in 1947; in Achimota, near Accra, Ghana, in 1958; in New Delhi, India, in 1961; in Mexico City, Mexico, in 1963; in Bangkok, Thailand, in 1972/1973; in Melbourne, Australia, in 1980; in San Antonio, Texas, USA, in 1989; in Salvador da Bahia, Brazil, in 1996; and the latest in Athens, Greece, in 2005.

In the Christian religion, from the very early stages, there were several conflicts, like the iconoclastic controversy over Mary's immaculate conception. Nobody knew how Mary got pregnant while she was not a married woman. A non-married woman in those days had to be a virgin. It was the culture where Mary was born, and the Bible says, "Suppose a man marries a woman who must be a virgin, and husband sleep with her then finds she is not virgin, the woman must be taken to the door of her father's home. There the men of the town must stone her to death, for she has committed a shocking crime in Israel by being promiscuous while living in the parents' home. In this way, you will purge this evil from among you" (Deut. 22:20–21). Based on the Bible, the survival of Mary was in doubt unless a powerful leader of the time protected her and spread the word that the Holy Spirit had impregnated Mary. Catholic Mariology is based on the following four beliefs:

- *Perpetual virginity*: Mary was a virgin before the pregnancy, and she remained a virgin after the birth of Jesus Christ.

The Catholic Church, as well as Eastern and Oriental Orthodoxy, believe in perpetual virginity. Protestants deny her perpetual virginity and intercession.

- *Mother of God*: The Catholic Church, as well as Eastern and Oriental Orthodoxy, believe in the Mother of God.
- *The Immaculate Conception*: Catholics believe the Virgin Mary's conception was free from sin by the goodness of her son Jesus Christ. The Eastern Orthodox disagrees with the Catholic Church. The Immaculate Conception and that Mary was not sinless is denied by Protestants.
- *The Assumption of Mary into heaven*: The Catholic Church and Eastern and Oriental Orthodoxy believe that the Virgin Mary got into heaven at the end of her earthly life. Those churches believe in this dogma and celebrate August 15 as a holy day of obligation. Protestants deny the assumption of Mary into heaven.

The bottom line is that ecumenism did not unite the Catholics, Eastern Orthodox, and Protestants for the past two thousand years. The intention of modern ecumenism from 1910 to the present time was to fight against communism and spread the Christian faith in the developed countries and especially in East Asia, where they did not believe in God. The modern ecumenical movement was very successful in converting people to Christianity, mostly in the Philippines, South Korea, Singapore, Indonesia, Hong Kong, Malaysia, Vietnam, Sri Lanka, Taiwan, Myanmar, China, India, and Thailand. On the other hand, religion is fading in European countries and also in the United States.

Summary

Ecumenism, also called the ecumenical movement or ecumenical tendency, is the perception of one Church serving God in the world and is a reflection of a central teaching of the early Christian belief. The early Church had many conflicts that called for ecumenical declaration. Conflicts among Christians started in the second century. The conflicts never resolved up to this date.

Emperor Constantine I forced all the churches to agree on specific doctrine. The first real version of the Bible was approved by the First Council of Nicaea in 325 CE. Athanasius, the bishop of Alexandria, wrote the approved Bible in 367 CE. Before 367 CE, a printed Bible did not exist. Emperor Constantine I was not a Christian at the time the Bible was written. In 381 CE, he converted to Christianity to save the empire.

It is not clear where the modern ecumenical movement started. Some assume the 1910 World Missionary Conference in Edinburgh, Scotland, was the beginning of the ecumenical movement, and some believe that the Roman Catholic Church tried to reconcile with the Reformation churches that had separated over theological differences. The real purpose of this conference was that the imperialism started to expedite the Christian missionaries in a different part of the world and later take over by military forces.

The modern ecumenical movement was very successful in gaining converts to Christianity, mostly in the Philippines, South Korea, Singapore, Indonesia, Hong Kong, Malaysia, Vietnam, Sri Lanka, Taiwan, Myanmar, China, India, and Thailand. On the other hand, religion is fading in European countries and in the United States.

CONCLUSION

> Religion is the daughter of Hope and Fear, explaining to ignorance the nature of the Unknowable.
>
> —Ambrose Bierce

> Religion is an illusion and it derives its strength from the fact that it falls in with our instinctual desires.
>
> —Sigmund Freud

> Religion is excellent stuff for keeping common people quiet.
>
> —Napoleon Bonaparte

> Religion is something left over from the infancy of our intelligence; it will fade away as we adopt reason and science as our guidelines.
>
> —Bertrand Russell

Primitive religions have occupied about 90 percent of religious history. They were not part of an institution. They had no priest, no king, and they were not part of any empire. They believed in animism or spirits that controlled them, their environment, and the animals around them. This belief was from the fear of dying. They practiced the same way as Australian aborigines, African aborigines, Andaman Islanders, and North American Indians. Males and females were equally looking for food resources, and they were not dependent on each other. The oldest in the family or tribe was usually selected as their leader. Sometimes the leader was selected based on physical structure and the ability to hunt large animals.

A new era of agricultural revolution was warranted around 10,000 BC due to an increase in the human population, so the development of more sophisticated stone technology, improved knowledge of how to raise cattle and how to plant evolved. All this learning and technology allowed the communities to settle down to a new way of living in small established villages in fertile areas with predictable climate, mostly along rivers and waterways. This new era gave the right of ownership of land, livestock, tools, and, most of all, status to the leaders and priests, resulting in a better life for them and a worse life for followers who never had association with leaders and priests.

At the same time that culture and religions changed, fear of dying did not change. The powerful grabbed the most significant and fertile land, gradually controlling villages and hamlets, and created a system of government to protect the owners and the powerful. The priesthood instituted by the powerful separated hunter-gatherers from direct communication with their spirits and leaders. The priesthood claimed that they were the agents of the gods and supernatural and had direct contact with them.

Thousands of gods were created to meet the desire of the powerful and the priests. The number and positions of gods and goddesses in ancient Egypt, Mesopotamia, the subcontinent of India, and China have been discussed in previous chapters. Some of the created gods and goddesses were promoted and some demoted and killed, like Osiris and others. Among the gods promoted was Hapi, the god of the annual flooding of the Nile River. Hapi, the great god, provided fertile soil and plenty of food and crops for the survival of the Egyptians and, therefore, was greatly celebrated.

During this era, two myths that continue in a slightly different version in Abrahamic religions are worth mentioning: the myth of creation and the myth of the great flood.

The myth of creation is associated with Enki, the god of deep seas, who is one of the trio of Mesopotamian gods. Enki, with his sister Mami, created mortals, seven of whom were males, and seven of whom were females. These mortals were created by Mami from a mixture of clay and the flesh and blood of one perfect intelligent god that was sacrificed. Also included in this mixture was spittle from other great gods. This story is the first version of creation, and it later was modified by Abrahamic religions under Adam and Eve (Gen. 1:10,12,18,21,31).

As myth describes, 60 times 3,600 years had passed since the creation of mortals, and more than 2,000 gods were not happy at this time. At the council of the gods, they decided to eliminate the entire mortal population from the face of the earth. Atrahasis, who was half-mortal and half-god, was looking for his lord Enki's advice. Enki told Atrahasis to build a boat to save humanity and all living things. After building the ship, Atrahasis put aboard the seeds of all living things. The council of the gods devised a couple of plans, and none of which worked. Finally, they planned a flash flood to last for seven days and seven nights. The entire population of mortals and other living things were killed by this flood. After the flood ceased, the boat of Atrahasis came to rest atop Mount Nimush. He made a sacrifice of thanks to Enki, his beloved god. He made an offering for the rest of the gods and provided food for them. The gods gathered like a swarm over the offerings. This story is the first version of the flood and it was later modified by Abrahamic religions, replacing Atrahasis with Noah and his ark (Gen. 6–8).

On the Indian subcontinent, Hinduism may have had its roots in the civilization that developed long before writing was discovered. The Bhimbetka rock shelters date from 30,000 BC. These rock shelters are from a Mesolithic site near present-day Bhopal, in the Vindhya Mountains in the province of Madhya. They are part of an archaeological site that spans from the prehistoric period to the historic period.

Hindu religious culture shows evidence of what may have been a cult of a goddess and a bull. Female figurines commonly were used for worship. Bull figurines were used on many steatite seals. These small figurines were found in all parts of India and may have come from pre-Vedic civilizations. In Hinduism, there are many millions of gods and goddesses. Some of them were promoted, some were demoted, and some were even killed. Some of the gods or goddesses

Conclusion

served the kings and some of them fought against evil kings. This is discussed in chapter 16, section 3, Hare Krishna.

As described above, religions, gods, and goddesses were created by the powerful and the kings to protect their power, control the ordinary populations, and expand their territories. Hinduism is one of the oldest religions on the earth, and it has ruled the entire Indian subcontinent for thousands of years. All the myths and hymns of Hinduism were preserved orally. Hinduism developed on the caste system, which divides Hindus into four castes of very rigid hierarchical groups based on their duties. The priestly class is highest, the warrior class is second, the peasant class is third, and the servant class is the lowest.

Indians got tired of millions of gods and goddesses and the caste system. In response to this and the atrocities to the lowest class, Jainism and Buddhism emerged as nontheistic religions. These religions believed all humanity is equal. During the life of Buddha, Buddhism did not have the backing of any king or leader until the reign of Ashoka the Great. During the reign of Ashoka the Great, Buddhism as religion was spread by the action of the sword all over the Indian subcontinent and beyond. Later, during the reign of Kanishka the Great, Buddhism was one of the most robust religions globally, accomplished by invading more lands and killing more people. See chart 1 and chart 2 for the rise and fall of world religions.

The war over power never stops. After the collapse of Gauda, the last king from the Gupta dynasty, and the coming of the Chero and Sena dynasties, Buddhism was wiped out and Hinduism came back. Buddhism did not have a god to direct them for the elimination of Hinduism. With the military power of Ashoka the Great and Kanishka the Great, a massive number of innocent people were killed. The same happened for the resurgence of Hinduism. All this happened on the Indian subcontinent.

In East Asia, both major religions, Confucianism and Taoism, did not believe in any gods or goddesses. Both Confucius and Tao began as philosophers and many years later, after their death, their followers changed to religious overtones. The power of Chinese emperors spread both religions. See chart 1 and chart 2 for the rise and fall of world religions.

In Southwest and Central Asia, people were tired of thousands of gods and goddesses who did not help other than killing innocent people by those gods' agents and kings. The new idea of one God was introduced by Zoroaster (Persian: Zarathushtra). Based on archaeological evidence and linguistic comparison, modern scholars believe that Zoroaster must have lived sometime between 1500 and 1200 BC. Zoroaster was one of the most exceptional and creative leaders and reformers. He was not a violent leader. He did not create any cults to force people to accept his belief. He was a reformer, and thought he could bring social order. Before his death, he introduced his beliefs and ideas to King Vishtaspa and his wife. They spread Zoroastrianism, which later became the official religion of the Persian Empire under Cyrus the Great. Zoroastrianism, by force, sword, and killing of innocent people, spread to Southwest and Central Asia as the largest religion of its time. See chart 1, chart 2, and schedule 1.[152]

Judaism is based on stories and tales. Their leaders were mythological characters. There is no known archaeological or independent research confirming the existence of Abraham. The word YHWH (Yahweh) was not an invention of Moses. Yahweh was already invented by Zoroaster and introduced into the Torah by the influence of Cyrus the Great. Cyrus the Great's name is mentioned and often praised in the Torah.

Moses was called a prophet, but his existence has never been verified by archaeologists. According to the myth, he rescued his people from slavery, and they wandered for forty years in the desert. Yahweh did not help him get to the promised land of Israel. Moses is described in detail in chapter 6. The statement of the promised land and God's chosen people were a massive setback to Jewish expansion and progress until the nineteenth-century reformers started to change their views. Other than the Bible stories, there is not a shred of archaeological or independent research confirming Moses's existence.

Christianity sprang from Judaism; it is the first reformation of Judaism. The Roman Empire was involved from the beginning in all affairs of Christianity, Judaism, and all religions of the empire. The Roman Empire destroyed Jerusalem in 70 CE. They were aware of the importance of religion to control other people. The Romans

Conclusion

believed that they could not succeed with Judaic beliefs of the promised land and God's chosen people. In Judaism, there is an element of racism. They reformed Judaism to a new religion of Christianity that was much more acceptable to others, and therefore spread faster in Europe. Christianity's holy book was limited only to the gospel until 367 BC. Some fragments were written in Greek between 45 CE and 140 CE, which were not used by ordinary people or churches. The gospel authors were Mathew, Mark, Luke, and John. Some of the authors, like Mark, were anonymous and can never be said to exist.

The First Council of Nicaea in 325 CE was ordered by Constantine the Great, who was not Christian, to come up with a clear vision, unity, and idea of Christianity. There was no God and no Jesus Christ to create Christianity other than the Roman Empire, whose goal was to control the world. The Roman Empire was strong and was able to expand Christianity throughout history. See chart 1, chart 2, and schedule 1.

Islam was founded by Muhammad on the Arabian Peninsula. In the west and east of this peninsula were two significant empires: Persian and Roman. This area was relatively poor, with a lack of water sources and poor pastureland. This created a rough tribal culture. Tribes were enemies of one another and a tribe's goats, sheep, camels, horses, slaves, women, and children were under permanent threat.

Muhammad came up with an idea of controlling Mecca, and he created a cult with a tribal culture that included bounty raiding and looting. He faced considerable resistance from his tribe and was forced to leave Mecca.

Based on the definition of Abrahamic religion, God is omnipotent, omniscient, omnipresent, and omnibenevolent. If this God was true, he would guide people without shedding a drop of blood. Muhammad was the one who created a God in his mind and used fictitious revelations from this God as needed to control his people. This is discussed in greater detail in chapter 13 under "Muhammad."

The third leader of the cult, Umar ibn al-Khattab, came to power, and although the Persian Empire was internally chaotic and weak, it was also vast, so it took Umar ibn al-Khattab many attempts to conquer the entire empire. He killed hundreds of thousands of

Zoroastrians with maximum ferocity and burned to ashes the largest libraries in Persia. See chart 1 and schedule 1.

The Islam and Christian empires controlled the world with great atrocity. Many other religions arose and could not stand up against the two evil empires. For example, Sikhism emerged to denounce the caste system of Hinduism and the converting of Hindus to Islam by the force of the sword.

At the end of the nineteenth century, colonialism flooded East, Southeast, and South Asia with Christian mercenaries and converted Shintoists, Taoists, Buddhists, Hindus, and Confucians to Christianity.

On the other hand, religion and God are fading in European countries and even in the United States and other parts of the world. See chart 2 and schedule 1.

CHART 1

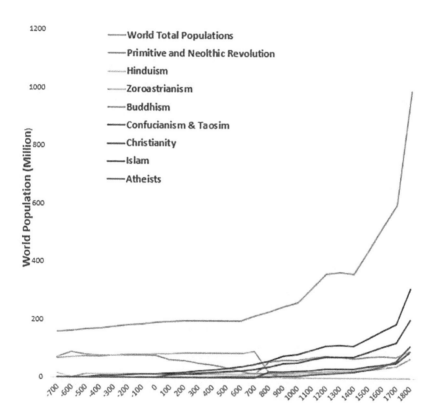

CHART 2

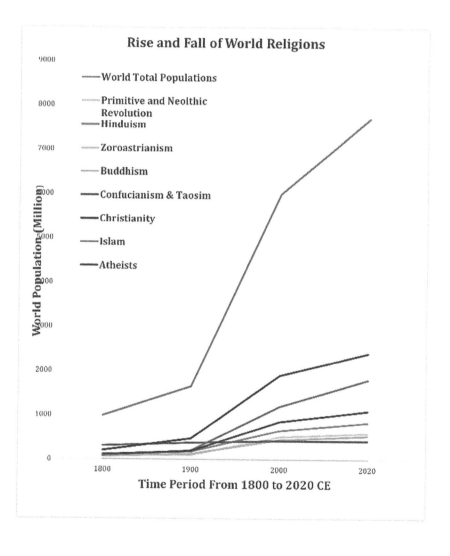

SCHEDULE 1

Year	World Total Populations	Primitive and Neolthic Revolution	RELIGIONS						
			Hinduism	Zoroastrianism	Buddhism	Confucianism & Taosim	Christianity	Islam	Atheists
-700	158	72	17	69	0	0	0	0	0
-600	163	90	1	71	0	0	0	0	0
-500	167	81	2	73	12	0	0	0	0
-400	172	76	2	75	12	7	0	0	1
-300	177	77	2	77	12	8	0	0	1
-200	181	77	2	79	13	9	0	0	1
-100	186	77	3	81	13	11	0	0	1
0	191	76	3	83	13	14	0	0	1
100	193	63	4	84	13	16	11	0	1
200	196	59	4	86	14	20	13	0	2
300	196	52	4	86	14	23	15	0	2
400	196	44	5	86	14	27	18	0	2
500	196	36	5	86	14	32	21	0	3
600	196	25	6	86	14	38	25	0	3
700	213	14	7	93	15	48	32	0	4
800	229	57	9	20	16	61	40	21	5
900	246	62	10	0	17	77	51	23	7
1000	262	63	12	0	18	82	54	24	8
1100	311	72	16	0	22	97	64	29	11
1200	360	79	21	0	25	113	74	33	16
1300	365	74	24	0	25	114	75	34	19
1400	360	68	26	0	25	113	74	33	21
1500	443	75	35	0	31	139	91	41	31
1600	525	78	46	0	37	164	108	48	43
1700	600	74	59	0	42	188	124	55	58
1800	990	95	108	0	69	310	204	91	113
1900	1650	100	180	0	127	380	474	200	189
2000	6000	512	655	0	448	428	1900	1200	857
2010	7000	721	764	0	483	432	2100	1500	1000
2020	7700	596	840	0.2	540	424	2400	1800	1100

BIBLIOGRAPHY

Akkermans, Peter M. M. G., and Glenn M. Schwartz. *The Archaeology of Syria*. Cambridge University Press, 2004.

Ali, Ahmed. *Al-Koran: A Contemporary Translation*. Princeton University Press, 2001.

Allen, John L., Jr. *The Future Church: How Ten Trends Are Revolutionizing the Catholic Church*. Image, 2012.

Ameer Ali, Syed. *Islam*. Cosimo Classics, 2019.

Ancient History Cyclopedia. www.ancient.eu.

Armstrong, Karen. *A History of God*. Ballantine Books, 1993

Armstrong, Karen. *Islam: A Short History*. Revised, updated, subsequent edition. Modern Library, 2007.

Armstrong, Karen. *The Battle for God*. Ballantine Books, 2001.

Aslan, Reza. *God: A Human History*. Penguin Random House, 2017.

Aslan, Reza. *No God but God: The Origins, Evolution, and Future of Islam*. Later print edition. Random House, 2005.

Aston, W. G. C.M.G. D Lit. *Shinto: The Ancient Religion of Japan*. Constable & Company Ltd, 1921.

Aston, W. G. *Shinto: The Ancient religion of Japan*. CreateSpace Independent Publishing Platform, 2015.

Baggini, Julian. *Atheism: A Very Short Introduction*. Oxford University Press, 2003.

Bailey, Cyril. *The Religion of Ancient Rome*. Kessinger Legacy Reprints, 2010.

Bankston, John. *Ancient India Maurya Empire*. Mitchell Lane Publishers, 2012.

Barnett, L. D. *Hinduism*. Forgotten Books, 2012.

Bartlett, Anne. *The Aboriginal Peoples of Australia*. Lerner Pub Group, 2001.

Bibliography

Bass, Diana Butler. *Christianity After Religion*. Reprint edition, Harper One, 2013.

BBC Religion and Culture. www.bbc.com

Beck, Glen. *It Is about Islam*. Threshold Editions, 2015.

Beery, Itzhak. *The Gift of Shamanism*. Destiny Books, 2015.

Berthrong, John H. and Evelyn Nagai Berthrong. *Confucianism: A Short Introduction*. Oneworld Publications, 2000.

Bertman, Stephen, PhD. *The Eight Pillars of Greek Wisdom*. Fall River Press, (2003) 2007.

Bettany, George Thomas. *World's Religions*. London: Ward, Lock & Co, 1890.

Bhaskarananda, *Swami. The Essentials of Hinduism*. Viveka Pr, 1995.

Bissell, Tom. *Apostle*. Reprint edition. Vintage, 2017.

Boyce, Mary. *Zoroastrians: Their Religious Beliefs and Practices*. Routledge, 1979.

Boyett, Jason. *12 Major World Religions*. Zephyros Press, 2016.

Branden, Charles Samuel, PhD. *The World's Religions*. Abingdon Press, 1954.

Breasted, James H. *Development of Religion and Thought in Ancient Egypt*. Cosimo Classics, 2010.

Brundle, Harriet. *Buddhism*. Booklife, 2017.

Burton, Jennifer. *Sikhism*. Mason Crest, 2017.

Carroll, Sean. *The Big Picture*. Dutton, 2016.

Claassen, William. *Alone in Community: Journey into Monastic Life Around the World*. Forest of Peace Books, 2000.

Cleary, Thomas. *The Taoism Reader*. Translated and edited by Thomas Cleary. Shambhala, 2012.

Cliteur, Paul. *The Secular Outlook*. Wiley-Blackwell, 2010.

Clodd, Edward. *Animism: The Seed of Religion*. Reprinted. CreateSpace, North Charleston, 2012.

Coleman, Cassie. *Hinduism: A Comprehensive Guide to the Hindu Religion*. CreateSpace Independent Publishing Platform, 2017.

Cox Harvey. *Fire from Heaven*. Da Capo Press, 2001.

Crossan, John Dominic, and Jonathan L. Reed. *Excavating Jesus*. Reprint edition. HarperCollins, 2003.

Dancing Genes Discovered. https://www.israel21c.org/dancing-genes-discovered-by-israeli-researcher/.

Darwin, Charles. *On the Origin of Species*. Barnes & Noble Classics, 2004.

Darwin, Charles. *The Indelible Stamp. The Evolution of an Idea*. Edited with commentary by James D. Watson. Running Press, 2005.

Dawkins, Richard. *The God Delusion*. Reprint edition. Mariner Books, 2008.

DeWitt, Jerry with Ethan Brown. *Hope After Faith*. Da Capo Press, 2013.

Doniger, Wendy. *The Hindus an Alternative History*. Penguin, 2010.

Donner, Fred M., et al. *The Oxford History of Islam*. Edited by John L. Esposito. Oxford University Press, 1999.

Dundas, Paul. *The Jains*. Psychology Press, 2002.

Eckel, Malcolm David. *Buddhism: Origin, Beliefs, Practices, Holy Texts, and Sacred Places*. Oxford University Press, 2002.

Ehrman, Bart D. *Did Jesus Exist?* Harper One, 2013.

Eliade, Mircea. *Shamanism Archaic Techniques of Ecstasy*. Later reprint edition. Princeton University Press, 2004.

Encyclopedia Britannica. www.britannica.com

Encylopedia.com. www.Encylopedia.com.

Epstein, Greg M. *Good Without God: What a Billion Nonreligious People Do Believe*. S. J. Coontz Company, 2005.

Evolution News and Science Today. www.evolutionnews.org

Feser, Edward. *The Last Superstition: A Refutation of the New Atheism*. St. Augustine Press, 2010.

Fohr, Sherry. *Jainism: A Guide for the Perplexed*. Bloomsbury Academic, 2015.

Fuentes, Agustin. *Why We Believe*. Yale University Press, 2019.

Gach, Gary. *Buddhism: The Complete Idiot's Guide*. Third edition. Alpha, 2009.

Ganeri, Anita. *Sacred Texts: The Guru Granth Sahib and Sikhism*. Smart Apple Media, 2003.

Ganeri, Anita. *World of Beliefs Buddhism*. Grange Books PLC, 2004.

Garcia, Mario T., and Vigil Elizondo. *The Gospel of Cesar Chavez: My Faith in Action*. Edited and introduced by Mario T. Garcia and Vigil Elizondo. Sheed & Ward, 2007.

Gardner, Daniel K. *Confucianism: A Very Short Introduction*. Oxford University Press, 2014.

Gibran, Kahlil. *The Prophet*. Alfred A Knopf, 1961.

Girzone, Joseph F. *My Struggle with Faith*. Reprint edition. Image, 2007.

Gray John. *Seven Types of Atheism*. Farrar, Straus, and Giroux, 2018.

Greenblatt, Stephen. *The Swerve: How the World Became Modern*. Norton Paperback, 2012.

Greene, Joshua. *Moral Tribe*. Penguin Books, 2014.

Griffith, Meghan. *Free Will: The Basics*. Routledge, 2013.

Griffiths, Bede. *The Cosmic Revelation: The Hindu Way to God*. Templegate Pub, 1983.

Haddon, Alfred C. *Magic and Fetishism*. Leopold Classic Library, 2015.

Hanson, Rick PhD. *Just ONE thing: Developing a Buddha Brain One Simple Practice at a Time*. Newharbinger Publication Inc., 2011.

Harrison, Jane Ellen. *The Religion of Ancient Greece*. Forgotten Books, 2012.

Hartz, Paula R. *Shinto World Religions*. Third edition. Chelsea House Publications, 2009.

Hartz, Paula R. *Taoism World Religions*. Updated edition. Facts on File, 2004.

Hatcher, William S., and J. Douglas Martin. *The Bahai Faith: The Emerging Global Religion*. Bahai Publishing, 2012.

Hazleton, Lesley. *Agnostic: A Spirited Manifesto*. Reprint edition. Riverhead Books, 2017.

Herman, Jonathan, PhD. *Taoism: For Dummies*. John Willey & Sons Canada, Ltd., 2013.

Hill, Jenna Miscavige, and Lisa Pulitzer. *My Secret Life Inside Scientology and My Harrowing Escape Beyond Belief*. William Morrow Paperbacks, 2013.

Hinnells, John R. *World's Living Religions*. Edited John R. Hinnells. Penguin Books, 2010.

History of the World. www.historyworld.net

Hoobler, Thomas, and Dorothy Hoobler. *Confucianism World Religions*. Third edition. Chelsea House Pub, 2009.

Hubbard, L. Ron. *The Scientology Handbook*. Bridge Publications, Inc., 1994.

Hulthkrantz, Ake. *Shamanic Healing and Ritual Drama, Health and Medicine in Native North American Religious Traditions*. Crossroad Publishing Company, 1997.

Husain, Ed. *The House of Islam, A Global History*. Bloomsbury Publishing, 2018.

Internet Sacred Text Archive. www.sacred-texts.com

Ions, Veronica. *Egyptian Mythology. Second Impression*. The Hamlyn Publishing Group Ltd. 1973.

Isaacson, Rupert. *The Healing Land (A Kalahari Journey)*. London: Fourth State, 2001.

Jain, Manoj, and Demi. *Mahavira the Hero of Nonviolence*. Wisdom Tales, 2014.

Johnson Linda. *Hinduism*. Second edition. Alpha Books, 2009.

Johnson, Paul. *A History of the Jews*. Harper Perennial, 1988.

Johsi, S. T. *The Unbelievers: The Evolution of Modern Atheism*. Prometheus, 2011.

Jordan, Michael. *Eastern Wisdom: Hinduism, Buddhism, Jainism, Taoism, Confucianism, Shinto, and Zen*. Carlton Books Ltd, 2003.

Kaltner, John. *Islam: What Non-Muslim Should Know*. Fortress Press, 2003.

Kanitkar, V.P., and Cole W. Owen. *Hinduism: An Introduction*. John Murray Press, 2011.

Kavelin, Chris. *Nudges from Grandfather, Honoring Indigenous Spiritual Technologies*. Christopher Kavelin, 2016.

Keown, Damien. *Buddhism: A Very Short Introduction*. OUP Oxford, 2013.

Kirk, Connie Ann. *The Mohawks of North America*. Lerner Pub Group, 2001.

Konner, Melvin, MD. *Believers: Faith in Human Nature*. W. W. Norton & Company, 2019.

Bibliography

Kriwaczek, Paul. *In Search of Zarathustra: The First Prophet and Ideas That Changed the World.* Orion Pub Co, 2003.

Krojeve, Alexandere. *Atheism.* Translated by Jeff Love. Columbia University Press, 2018.

LeDrew, Stephen. *The Evolution of Atheism.* Oxford University Press, 2015.

Lee, Helene, trans., and Stephen Davis, ed. *The First Rasta: Leonard Howell and the Rise of Rastafarianism.* Lawrence Hill Books, 2004.

Long, Jeffery D. *An Introduction: Jainism.* I. B. Tauris, 2009.

Lopez, Donald S. Jr. *Buddhism: The Norton Anthology of World Religions.* W. W. Norton & Company, 2014.

MacCulloch, Diarmaid. *Christianity: The First Three Thousand Years.* Reprint ed. Penguin Books, 2011.

Marisco, Katie. *Sikhism Global Citizens: World Religions.* Cherry Lake Publishing, 2017.

Martin, Michael. *Atheism: A Philosophical Justification.* Temple University Press, 1992.

McManning, John. *The Oxford Illustrated History of Christianity.* Edited by John McManning, Oxford University Press, 2001.

McQuail, Lisa (1960). *The Masai of Africa.* Singapore: Times Editions, 2002.

Melton, J. Gordon. *The Church of Scientology.* Signature Books, 2000.

Menzies, Robert P. *Pentecost.* Gospel Publishing House, 2013.

Meredith, Susan, and Clare Hickman. *Encyclopedia of World Religions.* Edited by Kirsteen Rogers. Edc Pub, 2002.

Moses, Jeffrey. *Oneness: Great Principles Shared by All Religions.* The Random House Publishing Group, 2002.

Myrtle Langley. *Religion.* DK Children, 2000.

Nardo, Don. *World Religions Buddhism.* Compass Point Books, 2009.

Nesbitt, Eleanor. *Sikhism: A Very Short Introduction.* Oxford University Press, 2005.

Neusner, Jacob. *World Religions in America.* 3rd ed. Westminster John Knox Pr., 2003.

New Living Translation. *Holy Bible: New Living Translation*. Tyndale House, 2006.

New World Encyclopedia. www.newworldencyclopedia.org

O'Donnell James J. *The End of Traditional Religion Pagans and the Rise of Christianity*. Reprint ed. Eeco, 2016.

Oldstone-Moore, Jennifer. *Confucianism*. Oxford University Press, 2002.

Oldstone-Moore, Jennifer. *Taoism*. Oxford University Press, 2003.

Oldstone-Moore, Jennifer. *Understanding Taoism*. Watkins, 2011.

Ostler, Nicholas. *Empires of the Word, A Language History of the World*. Harper Collins, 2005.

Paulos, John Allen. *Irreligion: Why the Arguments for God Just Don't Add Up*. Hill and Wang, 2008.

Pearcey, Nancy. *Finding Truth 5 Principles for Unmasking Atheism, Secularism, and Other God Substitutes*. David C. Cook, 2015.

Peet, Rev Stephen D. *The Religious Beliefs and Traditions of the Aborigines of North America*. Victoria Institute, 1885.

Peters, Jane. *Hinduism*. CreateSpace Independent Publishing Platform, 2016.

Petrie, W. M. Flinders. *The Religion of Ancient Egypt*. Kessinger Publishing, LLC, 2003.

Phillips, Stephen H. *Classical Indian Metaphysics*. Motil Banarsidass, 1998.

Phythian-Adams, W. J. *Mithraism*. The Open Court Publishing Company, 1915.

Pickren, Wade E. *The Psychology Book*. Sterling, 2014.

Radcliffe-Brown, A. R. *The Andaman Islanders*. University of Calif. Libraries, 1922.

Radley, Gail. *Understanding Islam*. Essential Library, 2018.

Rankin, Aidan. *The Jain Path Ancient Wisdom for the West*. Mantra Books, 2006.

Regan, Michael. *Understanding Sikhism*. Essential Library, 2018.

Reitman, Janet. *Inside Scientology: The Story of American's Most Secretive Religion*. Mariner Books, 2013.

Renard, John, PhD. *The Handy Religion Answer Book*. 2nd ed. Visible Ink Press, 2012.

Bibliography

Renou, Louis. *Hinduism Great Religion of Modern Man.* Edited by Louis Renou. G. Braziller. First edition, 1962.

Rhys, David, and Thomas William. *Early Buddhism.* Forgotten Books, 2012.

Riordan, James, and Jenny Stow. *The Coming of Night a Yoruba Tale from West Africa.* Gardners Books, 2000.

Rosen, Steven J. *Essential Hinduism.* Praeger, 2006.

Rossel, Seymour. *The Essential Jewish Stories.* KTAV Publishing House, 2011.

Routledge Companion Encyclopedias. *The World's Religions.* Edited by Stewart Sutherland, Leslie Houlden, Peter Clarke and Friedhelm Hardy. Routledge Companion, 2011.

Rushdie, Salman. *The Satanic Verses.* Reprint edition. Random House Trade Paperbacks, 2008.

Sachs, Curt. *World History of Dance.* Norton Library, 1963.

Sawyer, Dana. *Living the World's Religion.* Fons Vitae, 2014.

Sayce, A.H. *Assyro-Babylonian Religion.* Fourth edition. London: Williams and Norgate, 1897.

Schomp, Virginia. *The Ancient Persians.* Benchmark Books, 2009.

Sexton, John. *Standing for Reason: The University in a Dogmatic Age.* Yale University Press, 2019.

Sharma, Arvind. *Classical Hindu Thought and Introduction.* Oxford University Press, 2001.

Schouler, Kenneth, PhD., and Susai Anthony. *The Everything Hinduism Book.* Everything, 2009.

Schouler, Kenneth, PhD. *Everything World's Religions Book.* Second edition. Adama Media, 2002.

Simpkins, C. Alexander Ph.D., and Annellen Simpkins Ph.D. *Simple Confucianism: A Guide to Living Virtuously.* Tuttle Publishing, 2001.

Sir Frazer, James George. *Fear of the Dead in Primitive Religion.* Biblo & Tannen Publishers, 1966.

Skilton, Andrew. *A Concise History of Buddhism.* Barnes & Noble, 2000.

Smith, Huston. *The World's Religions.* Harper Collins. 1958.

Smith, W. Ramsay. *Myths and Legends of the Australian Aborigines.* Dover Publications, 2003.

Stafford, Tim. Miracles: Bethany House Publishers, 2012.

Stausberg, Michael. *Zarathustra and Zoroastrianism.* Equinox Publishing Limited, 2008.

Stenger, Victor J. *God and the Multiverse.* Prometheus Books, 2014.

Stenger, Victor J. *The New Atheism: Taking a Stand for Science and Reason.* Prometheus, 2009.

Steyn, H.P. *The Bushmen of the Kalahari.* Wayland, 1985.

Stroup, Herbert. *Four Religious of Asia.* Harper, 1968.

Telushkin, Rabbi Joseph. *Jewish Literacy.* Revised edition. William Morrow, 2008.

Matlins, Stuart M., and Arthur J. Magida. *The Essential Religious Etiquette Handbook.* Skylight Paths Pub., 2013.

The Global Religious Landscape. https://www.pewforum.org/2012/12/18/global-religious-landscape-exec/

The Koran. Translated by J. M. Rodwell and with an introduction by Alan Jones. Phoenix Press, Orion Publishing Group, 2005.

Anderson David A., and Sankofa. *The Origin of Life on the Earth: An African Creation Myth.* Illustrated by Kathleen Atkins Wilson. Sights Production, 1991.

Hinnells, John R., ed. *The Penguin Handbook. World's Living Religions.* Penguin Books, 2010.

Underhill, Ruth Murray. *Red Man's Religion: Beliefs and Practices of the Indians North of Mexico.* University of Chicago Press, 1972.

Wade, Nicholas. *The Faith Instinct.* Penguin Books, 2009.

Walsh, Roger, MD, PhD. *The World of Shamanism.* Llewellyn Publications, 2007.

Walter, Philippe. *Christianity the Origin of a Pagan Religion.* First US edition. Inner Traditions, 2006.

Wangu, Madhu Bazaz. *Buddhism: World Religions.* Fourth edition. Chelsea House Publications, 2009.

Waterhouse, John W. *Zoroastrianism.* The Book Tree, 2006.

Watson, Galadriel. *Bushmen of Southern Africa.* Weigl Pub Inc., 2012.

Watts, Alan. *Taoism Way beyond Seeking.* Tuttle Publishing, 2001.

Weber, Max. *The Religion of India.* Munshirm M, 2000.

Whitmarsh, Tim. *Battling the Gods: Atheism in the Ancient World.* Vintage, 2016.
Wilhelm, Richard Baynes. *The I Ching or Book of Changes.* Princeton University Press, 1977.
Wong, Eva. *Taoism: An Essential Guide.* Shambhala, 2011.
World-Religion-Professor.com. www.world-religions-professor.com
Wright, Lawrence. *Going Clear: Scientology, Hollywood, & the Prison of Belief.* Vintage, 2013.
Wright, Robert. *The Evolution of GOD.* Little, Brown and Company, 2009.
Yamakage, Motohisa. *The Essence of Shinto.* Kodansha International, 2012.
Yao, Xinzhong. *An Introduction to Confucianism.* Cambridge University Press, 2012.

INDEX

A

Aboriginals 29
Abraham 107, 111
Abrahamic faith 16. *See* also Judaism, Christianity, and Islam
Adventists 226
Agnosticism 335
agricultural revolution 47. *See* also Neolithic revolution
Ahura Mazda (supreme god of the Zoroastrian faith) 90, 95
Al-Aqsa Mosque 11
Amen-Ra (Egyptian deity) 52
Amesha Spentas 89
Amish Mennonites 224
Amun (Egyptian deity) 50
Andamanese 34
Anglican Church 231
Animism 29, 34, 47, 195, 208, 347
Anubis (Egyptian deity) 50
Anu (Mesopotamian deity) 59
Anunnaki (group of deities) 64
Anzu (Mesopotamian deity) 63
Arya Samaj 72
Atheism 329, 334
  on morality 336
  types of 331

Atrahasis (servant) 60
Aurangzeb 71

B

Baha'i 297, 307
  beliefs 301
  denominations 300, 301
  festivals and holy days 305
  leaders 298, 299
  on the concept of creation 302
  rituals and customs 304, 305
  scriptures 300
Bahaism 130
Baptists 223
Barrett, Justin L. 23
Bes (Egyptian deity) 50
Biden, Joe 26
Blanchard, Caroline 17
Blanchard, Robert 17
Brahma 74
Branch Davidians 24
Buddha 147, 148
Buddhism 16, 129, 146
  beliefs 156, 159
  death and afterlife 161
  denominations 155

Index

disciples and leaders 149, 150
religious festivals and holidays 154, 155
rituals and customs 161, 163
scriptures 152
Buraq 11
Bushmen 31

C

caste system 82
Castro, Fidel 23
Catholic Church 212, 216, 221
Catholic Mariology 343
Cerlarius, Michael 220
Chango. *See* Shango
Christianity 16, 129, 211, 241
    beliefs 231
    deity 232
    denominations 220, 222, 224
    ethics 232
    festivals and holy days 235
    influence of Zoroastrianism 237
    leaders 212, 213, 215, 218
    on death and the afterlife 233
    on morality 239
    rituals and customs 234
    Roman Catholic Sacraments 220
    salvation 232, 233
    scriptures 218, 220
    the beginning 211
Christian Science 230
Church of Jesus Christ of Latter-Day Saints, The 225

Clement VII (pope) 231
Confucianism 16, 129, 166, 176, 180, 185, 191, 193, 195, 201, 240, 350
    beliefs 175
    brief history in ancient China 166
    denominations 174
    festivals and holidays 177, 178
    leaders 170
    rituals and customs 176, 177
    scriptures 172, 173
Confucius 167, 168, 170
Creek Indians 40
Cruz, Ted 26
Cyrus the Great 102

D

Dakotas of Minnesota 42
Daoism. *See* Taoism
Dawkins, Richard 16, 331
Dayananda Saraswati 72
deities. *See* individual names of deities
Dennett, Daniel 331
Djhuty. *See* Thoth
Dreamtime 30

E

Eastern Orthodoxy 221
Ebstein, Richard 19
Ecumenism 339, 344
    ecumenical councils 339, 341
Edwards, Emanuel Charles (prince) 320

Egyptian mythology
    death and afterlife rituals 54
embalmment 56
Enki (Mesopotamian deity) 61
Enlil (Mesopotamian deity) 59, 61

G

Gandhi, Mahatma 26
Gandhi, Mohandas Karamchand 72
Gardner, Howard 22
Geshtu-e (deity) 59
Ghaznavid Empire 71
Ghorid Empire 71
Greek Empire 246
Gupta Empire 69

H

Hare Krishna 314, 316, 318, 326
Harris, Sam 331
Hathor (Egyptian deity) 50
Hawley, Josh 26
Healing Land, The (Isaacson) 33
heaven 243, 281
hell 243, 281
Hinduism 66, 85, 128
    beliefs 79
    ceremonies and festivals 84
    deities 73
    denominations 78
    Hindu reformers 72
    history and founding date 66
    leaders 75
    on karma and rebirth 80
    on rituals and customs 82
    on the soul 81
    scriptures 76
    the four stages of life and their rituals 81
    tradition 69
Hitchens, Christopher 332
Hitler, Adolf 23
Honorius I (pope) 340
Horus (Egyptian deity) 51
Hubbard, L. Ron 310
Humanism 335
Huxley, Thomas Henry 335

I

Igigi (the group of gods and goddesses) 61
Iroquois 39
Isaacson, Rupert
    The Healing Land 33
Ishkur (Mesopotamian deity) 63
Isis (Egyptian deity) 51
Islam 10, 16, 130, 245, 280
    Battle of Badr 252
    beliefs 270
    deity 270
    denominations 268, 270
    festivals and holy days 278
    five pillars 271
    leaders 247, 249, 258, 262
    on the afterlife 274
    on the concept of creation 273
    prophets 270
    rituals and customs 274, 276
    scriptures 264, 271
    six major beliefs 272
Islamic Empire 89

Isra (Journey of the Night) 10

J

Jainism 16, 129, 133, 143
    belief 138
    death and afterlife 139
    denomination 136, 137
    festivals and holidays 140
    five great vows 140
    leaders 134, 135
    rituals and customs 141
    scriptures 135
    temples 142
Jehovah's Witnesses 228
Johnson, Ron 26
Judaism 16, 107, 129, 130, 131
    beliefs 121, 122
    common beliefs 108, 109
    denominations 120
    division in the ancient era 121
    festivals and holy days 123, 124
    in the modern era 127
    leaders 110, 112, 114, 116, 118
    on morality 128
    rituals and customs 125
    scriptures 118
    temple 126

K

Kaepernick, Colin 24
Khnum-Ra (Egyptian deity) 52
King, Martin Luther Jr. 26
Koran 264, 265, 275. *See* also scriptures under Islam
    Meccan Phase 265, 267
    Medinan Phase 280
Koresh, David 23, 24

L

Lao Tzu 181
Leakey, Louis Seymour Bazett 28
Leo I (pope) 340
Leo IX (pope) 220
Leon X (pope) 217
Lutheranism 223

M

Mami (womb goddess) 64
Manah, Vohu 89
Mandela, Nelson 26
Martin V (pope) 217
Mauryan Empire 139
Mesopotamia 57
    religion in 58, 60
Methodist 224
Miller, William 226
Miraj (Muhammad's miraculous ascension from Jerusalem to the seventh heaven) 10
Mobbs, Dean 17
Mogul Empire 71
Mormons. *See* Church of Jesus Christ of Latter-Day Saints, The
Mount Hira 10
Muhammad 10, 245

N

Nagarjuna 150
Neolithic revolution 47
    religious practices 166

newer religions 323, 326, 327.
   See also modern under
   religions
Nintu 63
Ninurta (Mesopotamian deity)
   62
Nootkas 41
Nut (Egyptian deity) 51

O

Olodumare 34
Orthodox Brahmanism 159
Oshun 34
Osiris (Egyptian deity) 51
Osun. See Oshun
Oxus River 88, 97, 258

P

Parham, Charles Fox 313
Pence, Mike 25
Pentecostalism 312, 326
   festivals 314
Persian Empire 88, 89, 101,
   102, 103, 237, 254, 257
popes. See names of individual
   popes
Protestantism 222
Protestant Reformation 222
Ptah (god of Memphis) 52
Purushartha 79

Q

Quabootze 41
Quakers. See Society of Friends

R

Radcliff-Brown, A. R. 36

Radharani, Srimati 315
Ra-Horakhty (Egyptian deity)
   52
Ramakrishna, Paramahamsa 72
Rastafarianism 318, 320, 327
   beliefs 321, 322
   Youth Black Faith 320
Re (Egyptian deity) 52
Reformed and Presbyterians 224
religions 9, 12, 16, 47, 72, 73.
   See also specific names of
   religions
   Abrahamic 98, 102, 108,
      109, 119, 121
   cultural evolution 18
   four major 57
   from the beginning of
      Neolithic revolution
      48
   Middle Eastern 107
   modern 53
   primitive 20, 29, 44
religious symbolism 39
Rock Shelters of Bhimbetka 20
Roman Empire 49, 121, 126,
   211, 220, 242, 351
Russel, Charles Taze 228
Russell, Bertrand Arthur William
   17

S

Sasanian Empire 101
Sassanid Empire 94
Savarkar, V. D. 73
Scientology 310, 326
Selassie, Haile (emperor) 320
Set (Egyptian deity) 52
Seventh-day Adventist. See
   Adventists

Index

Shaivism 69
Shaktism 69
Shamanism 29, 323, 327
   beliefs 324
Shango 33
Shankara 70
Shintoism 195, 208
   beliefs 202
   denominations 201
   festivals and holidays 207, 208
   leaders 197, 200
   rituals and customs 203
   scriptures 200
   shrines 204
   worship 204
Shiva 74
Sikhism 130, 283, 295
   beliefs 288, 289
   denominations 288
   festivals and holy days 293
   leaders 283, 285
   on the afterlife 290
   rituals and customs 291
   scriptures 287
Sima Qian 181
Sindhu 66
Sobek-Ra (Egyptian deity) 52
Society of Friends 227
Stalin, Joseph 23

T

Tamil 70
Taoism 16, 129, 180, 193
   aspects of divinity 187
   beliefs 186, 187
   denominations 185
   festivals and holidays 191
   leaders 181, 182
   rituals and customs 189, 191
   scriptures 183, 184
Tao-Te-Ching 180
Temmu (emperor) 195
Temple, William 342
Theodosius I (emperor) 242
Thoth (Egyptian deity) 52
Torah 113
Tribes of Vancouver Island. *See* Nootkas
Trump, Donald J. 24

U

Unitarians 229
Universalism 229
Upanayana 84
Ur-Namma 58
Usir. *See* Osiris

V

Vaishnavism 69
Vatsyayana Mallanaga 80
Vigilius (pope) 340
Vishnu 74
Vivekananda, Swami 72

X

Xenophanes of Colophon 329

Y

Yang Zhu 180
Yoruba 33
Youth Black Faith 320. *See* also under Rastafarianism

Z

Zedong, Mao 23
Zoroastrianism 16, 88, 128
    beliefs 95
    deities 91
    denominations 95
    influence on Islamic religion 279
    influence on Judaism, Christianity, and Islam 102, 103, 105
    leaders 92
    on death 97
    on festival and holidays 98
    on final judgment 98
    on influencing Christian faith 238
    on rituals and customs 100, 101
    on the soul 96
    scriptures 93

NOTES

1. Robert Wright, *The Evolution of God*, 460–463.
2. https://www.livescience.com/21478-what-is-culture-definition-of-culture.html.
3. https://www.britannica.com/biography/Bertrand-Russell.
4. https://academic.oup.com/bioscience/article/50/10/861/233998.
5. https://pubmed.ncbi.nlm.nih.gov/14609538/.
6. https://humanorigins.si.edu/evidence/human-fossils/species/orrorin-tugenensis.
7. https://news.ucsc.edu/2013/04/sea-lion-beat.html.
8. https://theweek.com/articles/465272/6-animals-that-science-discovered-dance.
9. https://www.israel21c.org/dancing-genes-discovered-by-israeli-researcher/.
10. https://journals.plos.org/plosgenetics/article?id=10.1371/journal.pgen.0010042.
11. https://journals.plos.org/plosgenetics/article?id=10.1371/journal.pgen.0010042.
12. http://www.watarrkafoundation.org.au/blog/the-tradition-of-aboriginal-dance.
13. https://www.britannica.com/list/6-classical-dances-of-india.
14. https://theculturetrip.com/europe/turkey/articles/ancient-sufi-dance-rumis-whirling-dervishes/.
15. https://www.sciencedirect.com/science/article/pii/S0960982215014256.
16. https://www.sciencedirect.com/science/article/pii/S0960982215014256.
17. https://news.harvard.edu/gazette/story/2019/11/new-harvard-study-establishes-music-is-universal/.
18. https://www.bbc.com/future/article/20190529-do-humans-have-a-religion-instinct.
19. https://www.linguisticsociety.org/content/how-many-languages-are-there-world.
20. https://etc.ancient.eu/travel/visiting-ancient-city-kish/.
21. https://www.britannica.com/science/multiple-intelligences.
22. https://theness.com/neurologicablog/index.php/hyperactive-agency-detection/.
23. https://www.britannica.com/biography/Louis-Leakey.
24. https://www.britannica.com/biography/Edward-Burnett-Tylor.

Notes

25 http://www.perseus.tufts.edu/hopper/text?doc=Perseus:text:1999.01.0160:book=7:chapter=22.
26 https://www.britannica.com/biography/Pausanias-Greek-geographer.
27 https://www.britannica.com/topic/Black-Stone-of-Mecca.
28 https://www.britannica.com/topic/Australian-Aboriginal.
29 https://www.britannica.com/topic/Australian-Aboriginal.
30 https://www.britannica.com/topic/the-Dreaming-Australian-Aboriginal-mythology.
31 https://www.britannica.com/topic/Shango.
32 https://www.britannica.com/biography/A-R-Radcliffe-Brown.
33 https://www.livemint.com/Leisure/tIsiO3lUJFbVtlo39lIfIP/Getting-to-know-the-Andamanese.html.
34 https://www.timeanddate.com/time/us/arizona-no-dst.html.
35 George Thomas Bettany, *The World Religions* (London: Ward, Lock & Co. 1890), 64.
36 https://www.history.com/topics/ancient-history/ancient-egypt.
37 https://www.britannica.com/topic/Anubis.
38 https://www.britannica.com/topic/Isis-Egyptian-goddess.
39 https://www.britannica.com/list/11-egyptian-gods-and-goddesses.
40 https://www.britannica.com/list/11-egyptian-gods-and-goddesses.
41 https://www.britannica.com/topic/Busiris.
42 https://www.britannica.com/topic/Ptah.
43 https://www.britannica.com/topic/Re.
44 https://www.britannica.com/topic/Thoth.
45 http://anthropology.msu.edu/anp455-fs14/2014/10/28/duat/.
46 https://www.britannica.com/topic/Maat-Egyptian-goddess.
47 http://anthropology.msu.edu/anp455-fs14/2014/10/28/duat/.
48 https://www.britannica.com/topic/Lamashtu.
49 https://www.britannica.com/topic/Ea.
50 https://www.britannica.com/topic/Ishkur.
51 https://www.bbc.co.uk/religion/religions/hinduism/ataglance/glance.shtml.
52 https://www.britannica.com/topic/Shatapatha-Brahmana.
53 https://www.britannica.com/biography/Vinayak-Damodar-Savarkar.
54 http://www.bbc.co.uk/religion/religions/hinduism/deities/brahma.shtml.
55 http://www.bbc.co.uk/religion/religions/hinduism/texts/texts.shtml.
56 https://www.dummies.com/religion/hinduism/core-beliefs-of-hindus/.
57 https://www.britannica.com/topic/Purusha-Hindu-mythological-figure.
58 https://www.britannica.com/topic/samskara-Hindu-passage-rite.
59 https://www.britannica.com/topic/pumsavana.
60 https://www.wisdomlib.org/definition/simantonnayana.
61 https://jatakarma-hindu-birth-ritual.weebly.com/jatakarma.html.
62 https://www.weddingwire.in/wedding-tips/saptapadi--c2751.
63 https://www.britannica.com/topic/Holika.

64 https://www.britannica.com/place/Ayodhya.
65 https://www.britannica.com/topic/Ravana.
66 https://www.britannica.com/topic/Mahishasura.
67 Mary Boyce, *Zoroastrians: Their Believe and Practices* (Routledge, 1979).
68 http://www.heritageinstitute.com/zoroastrianism/zarathushtra/index.htm.
69 https://www.ancient.eu/Aryan/.
70 http://www.africansahara.org/ahriman-daevas-demons-zoroastrianism/.
71 Born 590 BC and died 529 BC.
72 http://www.heritageinstitute.com/zoroastrianism/scriptures/history.htm.
73 https://www.britannica.com/topic/Avesta-Zoroastrian-scripture.
74 https://www.britannica.com/topic/Bundahishn.
75 https://www.britannica.com/topic/Gayomart.
76 https://www.zoroastriankids.com/creation.html.
77 https://www.bbc.co.uk/religion/religions/zoroastrian/beliefs/god.shtml.
78 https://www.britannica.com/biography/Jacob-Hebrew-patriarch.
79 https://www.britannica.com/biography/Moses-Hebrew-prophet.
80 https://www.aish.com/sp/pg/48893292.html.
81 https://www.history.com/this-day-in-history/state-of-israel-proclaimed.
82 http://www.jewfaq.org/613.htm.
83 https://www.merriam-webster.com/dictionary/shivah.
84 https://kids.frontiersin.org/article/10.3389/frym.2016.00003.
85 https://www.irishtimes.com/culture/unthinkable/which-golden-rule-of-ethics-is-best-the-christian-or-confucian-1.1674003.
86 https://www.britannica.com/topic/Tirthankara.
87 https://www.britannica.com/biography/Parshvanatha.
88 https://www.britannica.com/topic/ratnatraya.
89 https://www.history.com/topics/religion/buddhism.
90 https://www.pewforum.org/2012/12/18/global-religious-landscape-exec/.
91 https://encyclopediaofbuddhism.org/wiki/Miracles_of_Gautama_Buddha.
92 https://www.learnreligions.com/the-buddhas-first-sermon-449788.
93 https://bigthink.com/ideafeed/good-news-science-buddha-agree-theres-no-you.
94 https://www.joincake.com/blog/confucianism-after-life/.
95 https://www.britannica.com/topic/ren.
96 https://www.ancient.eu/Lao-Tzu/.
97 https://www.britannica.com/biography/Yang-Zhu.
98 https://www.britannica.com/biography/Zhuangzi.
99 https://www.goodnet.org/articles/these-4-teachings-daoism-will-help-you-navigate-life.
100 www.unofficialroyalty.com/empress-teimei-of-japan-lady-sadako-kujo/.
101 https://www.tsunagujapan.com/10-japanese-traditional-rituals-to-give-every-child-a-happy-life/.
102 https://www.britannica.com/biography/Jesus.
103 https://www.britannica.com/list/st-pauls-contributions-to-the-new-testament.

Notes

104 https://www.pewforum.org/2012/12/18/global-religious-landscape-exec/.
105 https://www.churchrelevance.com/2012/06/22/qa-list-of-christian-denominations-and-their-beliefs/.
106 https://forums.catholic.com/t/why-is-it-that-protestants-do-not-believe-in-the-real-presence-its-in-the-bible/56547.
107 https://www.churchrelevance.com/2012/06/22/qa-list-of-christian-denominations-and-their-beliefs/.
108 https://www.churchrelevance.com/2012/06/22/qa-list-of-christian-denominations-and-their-beliefs/.
109 https://www.britannica.com/biography/Joseph-Smith-American-religious-leader-1805-1844.
110 https://www.britannica.com/biography/Joseph-Smith-American-religious-leader-1805-1844.
111 https://www.britannica.com/topic/Unitarianism.
112 https://answersingenesis.org/age-of-the-earth/how-old-is-the-earth/.
113 https://forums.catholic.com/t/why-is-it-that-protestants-do-not-believe-in-the-real-presence-its-in-the-bible/56547.
114 https://usqr.utsnyc.edu/wp-content/uploads/2011/08/JainUSQRv63-1-2.pdf.
115 https://www.pewforum.org/2012/12/18/global-religious-landscape-exec/.
116 https://www.britannica.com/topic/Quran.
117 https://www.answering-islam.org/Quran/Contra/.
118 https://www.britannica.com/topic/Adi-Granth-Sikh-sacred-scripture.
119 https://www.britannica.com/topic/gurdwara.
120 https://www.britannica.com/topic/Khalsa.
121 https://www.allaboutsikhs.com/scriptures/the-sikh-scriptures/.
122 https://www.merriam-webster.com/dictionary/halvah.
123 http://ranny-sunny.weebly.com/uploads/3/8/5/5/3855439/ceremony.pdf.
124 https://www.esplanade.com/offstage/arts/kirtan-singing-to-the-divine.
125 https://www.merriam-webster.com/dictionary/mela.
126 https://www.britannica.com/biography/Shoghi-Effendi-Rabbani.
127 https://www.pewforum.org/2012/12/18/global-religious-landscape-exec/.
128 https://fglaysher.com/bahaicensorship/ninedenominations.htm.
129 https://www.britannica.com/topic/qiblah.
130 https://bahaipedia.org/Laws.
131 https://study.com/academy/lesson/karl-lashley-theories-contributions-to-behaviorism.html.
132 https://tonyortega.org/2013/05/18/jon-atack-on-the-hypnotic-history-of-scientology-auditing/.
133 https://www.scientology.org/what-is-scientology/basic-principles-of-scientology/eight-dynamics.html.
134 https://www.scientology.org/store/item/science-of-survival-hardcover.html.
135 https://churchrelevance.com/qa-list-of-all-christian-denominations-and-their-beliefs/.

[136] https://www.apostolicarchives.com/articles/article/8801925/173171.htm.
[137] http://www.psychohistorian.org/display_article.php?id=200508010351_speaking_in_tongues.content.
[138] https://www.ancient.eu/Krishna/.
[139] https://www.history.com/topics/religion/history-of-rastafarianism.
[140] https://www.jstor.org/stable/1466399?seq=1.
[141] https://www.cambridge.org/core/journals/religious-studies/article/eliades-progressional-view-of-hierophanies1/C734DD4AECB39CC86F772F16A11EDB7F.
[142] https://www.merriam-webster.com/dictionary/scry.
[143] According to a 2015 Pew Research Center statistic.
[144] https://www.goodreads.com/quotes/8199-is-god-willing-to-prevent-evil-but-not-able-then.
[145] Richard Dawkins, *The God Delusion* (Leicester: W. F. Howes, 2007), 101, 102, 137, 151.
[146] https://rlp.hds.harvard.edu/religions/humanism/humanist-manifestos.
[147] https://americanhumanist.org/what-is-humanism/definition-of-humanism/.
[148] https://www.learnreligions.com/agnosticism-and-thomas-henry-huxley-248044.
[149] https://www.encyclopedia.com/philosophy-and-religion/philosophy/philosophy-terms-and-concepts/freethinkers.
[150] https://www.merriam-webster.com/dictionary/Origen.
[151] http://www.bu.edu/missiology/missionary-biography/t-u-v/temple-william-1881-1944/.
[152] https://worldpopulationreview.com.

ABOUT THE AUTHOR

Younus Samadzada was born in Kabul, Afghanistan but after graduating from high school, he traveled to the Czech Republic where he earned his degree in structural engineering. The Russian invasion of Afghanistan led him to the United States. He settled in Albany, New York, where he practiced engineering for over thirty-two years.

Growing up in a country that deeply valued religion and used it as an explanation to describe the world around him, he developed a keen interest in learning how religion shaped our thoughts and way of life. Over time, he grew more curious in the way religious belief systems developed, why they developed, and why the focus of many systems was the presence of a "God." At the age of fourteen, he made a commitment to himself that he would one day write a book that would take a deep dive into belief systems from the beginning of humans on earth to our present day.

CPSIA information can be obtained
at www.ICGtesting.com
Printed in the USA
FSHW010224110222
88215FS

9 781637 101414